# Revolting!

ALSO BY MICK HUME

*Trigger Warning*

# Revolting!

## How the Establishment Are Undermining Democracy and What They're Afraid Of

**Mick Hume**

WILLIAM
COLLINS

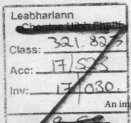

William Collins
An imprint of HarperCollins*Publishers*
1 London Bridge Street
London SE1 9GF
WilliamCollinsBooks.com

First published in Great Britain in 2017 by William Collins

1

Copyright © Mick Hume 2017

Mick Hume asserts the moral right to
be identified as the author of this work

A catalogue record for this book is
available from the British Library

ISBN 978-0-00-822082-2

Printed and bound in Great Britain by
Clays Ltd, St Ives plc

**MIX**
Paper from
responsible sources

FSC
www.fsc.org    **FSC** C007454

FSC™ is a non-profit international organisation established to promote
the responsible management of the world's forests. Products carrying the
FSC label are independently certified to assure consumers that they come
from forests that are managed to meet the social, economic and
ecological needs of present and future generations,
and other controlled sources.

Find out more about HarperCollins and the environment at
www.harpercollins.co.uk/green

For Ginny, my wife – one in 17.4 million

# Contents

# Author's note

This short book was conceived and written after the 2016 referendum on the UK's membership of the EU, and the US presidential election. But the idea of writing a polemical defence of popular democracy against its modern enemies did not occur to me overnight.

For some time, I had been aware of the widening gap between the rhetorical support for democracy in Western societies, and the attempts to restrict it in practice. And I had become particularly concerned by the Western left's effective abandonment of the defence of democratic freedoms.

Similar concerns motivated the writing of my previous book, *Trigger Warning: Is the Fear of Being Offensive Killing Free Speech?*, to which *Revolting!* can be seen as a companion, if not a sequel.

In contrast to today's state-centred leftism, the British left-wing tradition with which some of us identify has always taken the fight for democracy and free speech seriously: from the Levellers to the radical wing of the Suffragettes.

Those who believe in progress have fought for more democracy and freedom, not just as a good idea or an end in itself, but also as the means to help change the world. The left's abandonment of those historic causes marks the end of that era.

The need now is to bring the politics of democracy to life and start a new public debate about the sort of society we want. This book is intended as a contribution, and a call to arms.

*Revolting!*'s argument for more democracy has been developed through my years as a campaigning journalist in both the mainstream and alternative press, not least as the editor of *Living Marxism* magazine from 1988 (relaunched in the 1990s as the taboo-busting *LM* until it was forced to close after being sued under England's atrocious libel laws in 2000), and then as the launch editor of *Spiked* (spiked-online.com), of which I am now editor-at-large.

Like anything to do with mass democracy, however, this book is no solo effort. It would not have been possible without the contributions of others. I want to extend my thanks and admiration to my hard-working colleagues at *Spiked* – Tom Slater, Tim Black, Viv Regan and Ella Whelan, led by the inimitable editor Brendan O'Neill. They are true pioneers breaking new ground in the struggles over democracy and free speech.

Although this book is critical of the way that experts are now empowered to give orders rather than advice, it is itself reliant on the invaluable expertise of other writers and historians, past and present. I am indebted to the work of Bruno Waterfield and James Heartfield on the European Union.

Thanks are also due to Joseph Zigmond, my editor at William Collins, for helping to make the idea behind *Revolting!* into a reality. Finally, I would like to offer sincere thanks to my old friend and collaborator Frank Furedi, for the inspiration and advice to focus on the arguments that matter.

When it comes to taking responsibility for the text, warts and all, I remain of course in a minority of one.

Mick Hume, London, February 2017

# 1

# From Brexit to Trump: '... but some voters are more equal than others'

This is not a book about Brexit. Nor is it a book about the election of Donald Trump. It is about a much bigger issue – one the debate around those extraordinary events has highlighted. What's at stake is the future of democracy itself, in the UK, the US and across the West.

We live at a strange moment in the history of democratic politics. Today, perhaps for the first time, every serious politician and thinker in the Western world will declare their support for democracy in principle. Yet in practice the authorities are seeking to limit democratic decision-making and separate power from the people.

They invest authority instead in unelected institutions, from the courts to the European Commission. Elected politicians act as a professional elite, divorced from those they are supposed to represent. And everywhere, the intellectual fashion is to question whether voters are really fit or qualified to make democratic decisions on major issues, such as membership of the European Union or the Presidency of the United States.

Ours is the age of 'I'm a democrat, but ...', when the establishment insists it is all for democracy, but only in moderation; it just cannot tolerate what one former member of President

Barack Obama's administration calls 'too much of a good thing', suggesting that America 'might be a healthier democracy if it were a slightly less democratic one'.[1] For some in high places these days it seems that, where democracy is concerned, less really can be more.

It is an attitude captured in the UK by former Conservative prime minister John Major who, dismissing the suggestion that the Brexit referendum result should be binding, declared that 'the tyranny of the majority has never applied in a democracy'.[2] Some of us might naively have imagined that majority rule was the essence of democracy. But not, it seems, when millions of common 'tyrants' vote against the wishes of the minority political elite.

It is time we started not only to defend popular democracy, but to argue for far more of it, with no ifs, buts or by-your-leaves.

So this is not just a book about Brexit, or about Trump. The best place to begin the argument, however, is with the fallout from that UK referendum and the US election, which has brought the bigger picture sharply into focus. Whichever side you were on in those votes, the wider issue of your right to decide is now on the line.

In *Animal Farm*, his 1945 allegorical novel about the Soviet Union's descent from popular revolution to Stalinist tyranny, George Orwell gives one of the great definitions of the betrayal of democracy. It becomes clear that the farm has turned into a totalitarian system when the powerful elite of pigs alters the founding principle of Animal Farm painted on the barn wall. To the noble declaration 'All Animals Are Equal' they add the qualification '... But Some Animals Are More Equal Than Others'.[3]

Of course that's only dystopian fiction from 1945, in the faraway era of world war and totalitarianism. Even Orwell originally entitled it *Animal Farm: A Fairy Story*. It couldn't happen here.

Fast-forward to September 2016 and the director of the pollsters BritainThinks goes on BBC Radio 4's flagship morning *Today* programme, to discuss the findings of their focus-group conversations with voters from both sides of the June referendum on the UK's membership of the European Union.

Deborah Mattinson reported that some of the victorious Leave voters 'think the Remainers are rich people' who had benefited from the status quo within the EU. Something of a generalisation no doubt, but fair enough, perhaps. That argument at least acknowledges that those Remain voters had made a reasonable decision that might be seen as in their own self-interests. Think-tank research did find that, in the words of one Tory newspaper, 'Britain's ruling classes were the only group to vote overwhelmingly to stay in the European Union'.[4]

And what about the other side of the divide? What did the losing Remain voters in Mattinson's focus groups think of the opposite lot? Well, she said bluntly, 'Some of the Remainers think that some of the Leavers were stupid and shouldn't have the vote.' This revelation almost had *Today*'s world-weary host John Humphrys choking on his croissant.[5]

So it became possible, not in the allegorical dystopias of 1940s fairy tales but in the real world of twenty-first-century British politics, to hear it seriously proposed by some that some members of the electorate, though formally qualified to participate in our democracy, are 'stupid and shouldn't have the vote'. Or as Orwell's oligarchical pigs might have put it, 'All Voters Are Equal, But Some Voters Are More Equal Than Others.'

That report was no one-off. The 'too thick to vote' point might have been particularly blunt, but the underlying sentiment was the stuff of countless tweets, posts, articles, outbursts and reports in response to the referendum. The essential message was that all those Leave voters don't know what's good for them. The implication was that they should not have been allowed the right to make the wrong choice on such an issue.

That sneering attitude was even reflected in the satirical magazine *Private Eye*; under the spoof headline 'Turkeys Vote for Christmas in Referendum Cliffhanger', it reported that some turkeys were already regretting their 'Brexmas vote' as 'evidence is piling up that, come Christmas lunch, they will in fact have their heads cut off, their giblets put in a plastic bag and be well and truly stuffed'. If it was irony the *Eye* was after, how about 'Satirists Side with Establishment'?[6]

Then came the second political earthquake of 2016 – the November election of Republican candidate and celebrity capitalist Donald Trump as the forty-fifth president of the United States. The bitter responses to the voters' failure to elect Democratic Party favourite Hillary Clinton were if anything even more starkly anti-democratic than the anti-Brexit backlash.

'Your Vote is a Hate Crime!' declared anti-Trump protesters, graffiti artists and bloggers, implying that Trump supporters should be denied not only their vote, but their liberty.[7]

One leading Democrat commentator issued the blanket declaration that 'There's No Such Thing as a Good Trump Voter'. Meanwhile a celebrity professor of political science drew the no doubt scientific conclusion that Trump's victory was 'the dance of the dunces', made possible by wasting the right to vote on 'uneducated, low-information white people'.[8] Some voters, it appears, are now deemed 'more equal than others' because they are considered better-informed, or just better people.

As with the barn in *Animal Farm*, here too it appears that the writing is on the wall. The Brexit vote and the Trump election have shone the spotlight on democracy. Many in the upper reaches of politics, the media and culture do not like what they see.

They fear that they are witnessing a revolt of those whom candidate Clinton branded 'the deplorables' during the US election campaign. And they find the idea of such deplorable people exercising democratic power frankly revolting.

Reservations about allowing the people to vote and have some power over their lives have been around ever since the ancient Greeks invented the concept of democracy. As we explore in chapter 3, even in the modern era democracy was long considered a dirty word in the upper echelons of Western societies. It is only in more recent times that these prejudices have been restrained beneath the surface of polite society, as everybody has felt obliged to pay lip service to the principle of democracy.

But the fury of the political, economic and cultural elites in response to the 17.4 million UK voters who dared to back Brexit, and the 62 million-plus Americans who had the temerity to vote for Trump, brought these anti-democratic poisons bubbling to the surface of our civilised societies once more.

## The real Brexit–Trump connection

There has been a concerted attempt to explain the link between the Brexit referendum result and the election of Donald Trump. For angry social media commentators, it seemed obvious that 'both were clearly mired in racism, bigotry and hate'. Many mainstream media pundits took a similar line, concluding that 'both votes were marked by emotional, divisive campaigns' and were won on 'a tide' of racism and hate.[9]

Much of this misses the point. The important link between the Brexit and Trump votes was not the campaigns, but the reaction they provoked. Both results were met by an extraordinary outburst of fear and loathing from political and cultural elites, revealing their barely concealed contempt for the people and democracy. If there has been a dangerous hatred on view, it is the hatred of the 'herd' on both sides of the Atlantic.

To be clear from the start: while I supported the Brexit vote, I have no truck with Trump. The parallels are only in the way those backing the two campaigns have been condemned from on high.

Both results reflected the intensity of feeling against the respective political establishments. But the outcome was different. Whereas the vote to Leave the EU represented a positive blow for more democracy, the turnout for Trump was a negative reaction to the same problem of a political elite lacking legitimacy. There is a difference between supporting a broad democratic principle in a yes/no referendum, and backing a specific party's narrow-minded candidate in an election.

That is why some of us in the UK who voted Leave with passion could not have contemplated voting for the illiberal, free-speech-stomping Donald. Nor, by the way, could we have stomached supporting the illiberal-liberal Hillary Clinton. (Note to the confused: the Brexit referendum result was not a vote for Trump fan and UK Independence Party leader Nigel Farage, who responded to his triumph by giving up politics rather than taking power.)

No; the genuine comparison between the two concerns not the actors, but the anti-democratic reaction to the results. The backlash against Brexit set the pattern.

\* \* \*

6

On 23 June 2016, the British electorate went to the polls to vote in a referendum on whether the nation should Remain a member of the European Union, or Leave the EU. They voted to Leave, by 51.8 per cent to 48.2. The 17.4 million who voted Leave constituted the largest number of people who have ever voted for anything in British political history; the 16.1 million who backed Remain made up the second-largest vote for anything, reflecting the importance of the issue. (The most votes ever acquired by any party in a UK general election were the 14.1 million won by Conservative prime minister John Major in 1992 – representing 41.9 per cent of the votes cast. Oddly, Major did not seem to object to the tyranny of the minority on that occasion.)

The result was a remarkable popular rejection of the institutions of the EU – which, as chapter 4 argues, have been one of the major barriers to the practice of democratic politics in Europe. It represented a demand for more democracy and national sovereignty, and less diktat from the Euro-bureaucracy. It was also a sharp slap in the face for the British political class, who have long used the EU to sidestep democratic debate at home.

The UK's political, economic and cultural elites, who had all assumed until the last minute – along with every pollster, pundit and bookmaker – that Remain would win easily, reacted to the referendum result as if an earthquake had caused the solid ground to disappear from beneath their feet. How could this have happened?

After all, the Remain campaign had marshalled every authority in the Western world to warn those British voters that a Leave vote would lead to economic ruination, a political descent into barbarism, world war and, worse, falling house prices.

They had been told to vote Remain by the leaders of all Britain's mainstream political parties, from Tory prime minister David Cameron to left-wing Labour opposition leader Jeremy Corbyn. They had been instructed that there was no realistic alternative to voting Remain by the governor of the Bank of England, the Chancellor of Germany, the President of the United States, a cross-section of leading lights from the arts and every imaginable celebrity from David Beckham to Johnny Rotten. For its part, the official Leave campaign had looked like an embarrassing shambles. Yet still a majority of voters had refused to do as they were told they must, and opted to Leave the EU.

In the eyes of the establishment it appeared that the only possible explanation for this outrageous outcome was that those millions of voters were simply too ignorant, too uneducated, too gullible, bigoted or emotional to understand what they were being told. Leave voters were depicted as being like that naughty child whose finger is drawn inexorably towards the big, red button by all the warning signs telling him 'Danger – Do Not Press'.

Most striking was how quickly the discussion ceased to be about the specific issues of Brexit, and became about much bigger questions of democratic decision-making. The emphasis shifted away from what the electorate thought of the EU, towards what the pro-EU elites thought of the revolting electorate. Answer: not much. It may take a long time for the wrangling over the details of UK–EU relations to become clear. But the wider threat to democracy in the anti-Brexit backlash was evident from the start.

To clarify: this book's attack on the antics of the pro-EU elites is not aimed at the 16.1 million who voted to Remain. That would be a remarkably large 'elite' by anybody's standards.

Most of those Remainers were normal voters who made a rational choice, just as the Leavers did. Millions of them are also respecters of democracy. In a YouGov survey published in November 2016, 68 per cent of all respondents said that the UK should follow the referendum result and go ahead with Brexit. Those who had voted to Remain in June were now 'evenly divided' between those who 'think the government has a duty to implement the decision and leave' and those who 'would like to see the government ignore or overturn the referendum result'.[10]

The political, economic and cultural elites leading the anti-democratic campaign to 'ignore or overturn the referendum result' were a small minority within that minority, symbolised by such big-name, big-headed Remainers as Tony Blair or Sir Richard Branson. The 2016 poster girl for their crusade was Gina Miller, the multi-millionaire investment fund manager who led the legal challenge to the government over Brexit, because she said the revolting voters' verdict made her feel 'physically sick'. After the high court found in her favour, Ms Miller the City financier declared that the abuse she had received 'means I am doing something right for investors'.[11] This clique constitutes an elite in the worst sense of the word, defined by the *Oxford English Dictionary* as a class of people 'having the most power and influence in a society', not due to any superior talents but 'especially on account of their wealth or privilege'.

Weeks after the vote, European President Jean-Claude Juncker gave the official EU version of events in an interview with a French youth YouTube channel (where else would you make an important announcement these days?). Monsieur Juncker claimed that the blame lay with British politicians who had spent more than forty years spreading 'so many lies, so many half-truths' about the EU, telling 'your general public that European Union is stupid, that there is nothing worth ...'[12] His underlying message was that the 'general public' in the UK must have been sufficiently 'stupid' to believe whatever lies the politicians fed them.

Yet if anything looked ignorant or misinformed in this discussion, it was Juncker's claim that influential British politicians have been indoctrinating the general public with anti-EU 'lies and half-truths' for more than forty years.

Almost until the referendum campaign began, the political outlook labelled 'Euro-scepticism' had been a fringe affair, considered in parliament to be the preserve of only a few Tory head-bangers. Since the UK joined what was then the European Economic Community in 1973, no government had advocated leaving. The last time any major UK political party pledged to leave the EU at a general election was back in 1983, when it formed part of the Labour Party's left-wing manifesto – described as 'the longest suicide note in history' – which resulted in a devastating defeat.

In the June 2016 referendum campaign, the leaders of every mainstream party – including Labour's Corbyn, supposedly a long-standing left-wing Eurosceptic – backed the conformist Remain campaign. Even leading Tory Leave campaigner Boris Johnson had no history of being anti-EU, and had gone so far as to write an (unpublished) pro-Remain column months before the referendum.

The popular Brexit vote looked far more like a spirited revolt against discredited and two-faced politicians than any tame acquiescence to their instructions. In response, those politicians reacted as if they had been shot at. After the referendum Cameron quit as prime minister with a speed normally reserved for political leaders who are assassinated in office.

Bewildered leading Members of Parliament from all sides joined hands to bemoan the 'national disaster' of the Brexit vote. The immediate reaction was well captured by Labour MP and former government minister David Lammy, who tweeted a desperate appeal to his fellow members of the political class: 'Wake up. We do not have to do this. We can stop this madness and bring this nightmare to an end through a vote in Parliament … there should be a vote in Parliament next week.'[13] For the Right Honourable Lammy it seemed a display of popular democracy was madness, people voting other than as instructed a nightmare. All honourable parliamentarians needed to wake up and overturn the historic referendum result within the week.

Another senior Labour MP, Keith Vaz, bewailed the 'crushing, crushing decision … a terrible day for Britain … catastrophic. In a thousand years I would never have believed the British people would have voted in this way'. So how could a majority of those who voted – including his own constituents in Leicester – have done so in such an unbelievable, catastrophic fashion, and inflicted what Vaz seems to think was Britain's most terrible day since *circa* 1066? They voted, concluded Vaz, 'emotionally rather than looking at the facts'.[14] It couldn't possibly be that voters had looked at 'the facts' and reasonably drawn the opposite conclusion from their MPs; it had to be that the naughty children had let their feelings run away with them.

Politicians and lobbyists who claim to be most in favour of change in the UK seemed among those most upset by the popular vote to change Britain's relationship with the EU. Progressives and the Left have historically been the people who fought to 'leave' the current state of the world. Yet now they appeared determined to 'remain' in the status quo of the conformist EU.

The establishment's call for a Remain vote had been backed by leading liberal and left-wing voices from the *Guardian* to the *New Statesman*, the Labour Party mainstream to the 'Corbynite' Momentum campaign. Some reacted with bitterness and bile when the popular vote went against them. *Guardian* columnist Polly Toynbee, grand dame of British liberalism, denounced the 'stupidity' of the Leave campaign and demanded that 231 Labour MPs – 70 per cent of whose constituencies returned majorities for Leave – must be 'brave' and vote to 'save us' from the votes of 17.4 million Leavers – in the name of 'representative democracy', of couse.[15]

Nationalist politicians whose declared aim is to enable their people to break free from the United Kingdom appeared particularly furious at any suggestion that the British people should want to break free from the European Union.

First Minister Nicola Sturgeon of the Scottish National Party declared a state of national 'fury' over the Brexit vote (the majority of voters in Scotland supported Remain) and threatened to veto Brexit, in the apparent belief that democracy means 1.66 million Scottish Remain votes are so much more equal than others that they can outweigh 17.4 million Leave votes from across the UK.[16]

In the province of Northern Ireland, where a majority backed Remain, Sinn Fein's Martin McGuinness denounced the 'toxic' UK vote and declared that: 'The island of Ireland is facing the biggest constitutional crisis since partition [in 1921] as a result of the Brexit referendum.'[17] This might have come as a surprise to those who recall the 'constitutional crisis' posed by the twenty-five-year armed conflict over sovereignty that raged in Northern Ireland from 1969, which first brought Mr McGuinness to public attention. For this leading Irish republican, however, it appears that a popular vote for Britain to leave the EU is now far more 'toxic' than the arrival of British troops to keep Northern Ireland within the UK.

Elsewhere the Leave vote was dismissed by leading UK liberal writers as a 'howl of rage',[18] as if those voters had been little more than dumb animals responding like pups to the 'dog-whistle politics' of xenophobic demagogues; a modern reincarnation of the howling, foul-breathed 'beast with many heads', as Shakespeare's arrogant Roman general Coriolanus brands the people of Rome.

The consensus appeared to be that Leave voters must have taken leave of their senses to go against the advice of their betters. These responses let slip the mask and revealed the old elitist prejudices about the people not being fit for our democracy (rather than the other way around).

Like every leading anti-democrat since Plato, who wanted to replace the roughhouse of Athenian democracy with the rule of philosophers and experts, the political elites of the UK and Europe believe that matters of government are far too complex and sophisticated to let the governed decide. 'We all know what to do, we just don't know how to get re-elected after we've done it,' as EC President Juncker once said, in his previous life as prime minister of the Grand Duchy of Luxembourg (before becoming Duke of the Grand Duchy of Brussels).[19] Better by far, then, not to bother the masses' little heads with such democratic nonsense as elections and referendums wherever possible.

After the shock of the Brexit result, one might have expected the transatlantic elites to be ready for an upset in the coming US presidential election. Yet such was their smug complacency that they remained convinced the American people would take their instructions, reject the wild-talking maverick Donald Trump, and elect the respectable machine politician Hillary Clinton.

Less than a fortnight before polling day, a leading UK liberal commentator was berating the 'political and media class' for continuing to cover Trump's failing campaign rather than focusing on the real issue – the coming Clinton presidency: 'The big question in American politics is not whether Hillary Clinton will be president. It is what kind of president she is likely to be.'[20] On the eve of the election, the pollsters and bookmakers all seemed to agree that Clinton was a certainty for the White House.

When, on 8 November, the American electorate dared to disagree with these premature verdicts, and instead handed Trump the keys to the White House via the electoral college, there appeared to be even greater astonishment than after the Brexit referendum. How could this have happened?

After all, Trump had not only been denounced as a disgrace to US politics by the Democratic Party establishment, but also effectively disowned by all but a handful of senior figures from his own Republican side. The media too had been overwhelmingly anti-Trump, with only two established regional newspapers backing him in the entire United States.

And the worlds of Hollywood and celebrity, considered so influential in public life today, had been solidly for Hillary over Donald, staging a series of last-minute concert-rallies featuring the likes of Beyoncé and Jay-Z, Lady Gaga and Madonna, with a bit of Jon Bon Jovi and Bruce Springsteen thrown in for the wrinklier voters. How could Americans resist being dazzled by such a star-studded appeal?

When more than 62 million Americans did just that and voted for Trump, the reaction was a mixture of consternation and condemnation. Leading liberal voice Arianna Huffington declared the election of Trump to be simply 'incomprehensible'. After all, the blogging mega-site she founded, the Huffington

Post (still bearing her name though under different direction), had attached this editorial reminder to every report about the Trump campaign: 'Donald Trump regularly incites political violence and is a serial liar, rampant xenophobe, racist, misogynist and birther who has repeatedly pledged to ban all Muslims – 1.6 billion members of an entire religion – from entering the U.S.' Couldn't these 62 million people read?[21]

David Remnick, editor of Big Apple institution the *New Yorker*, immediately pronounced Trump's election to be not just incomprehensible but 'an American Tragedy ... a tragedy for the American republic, a tragedy for the Constitution, and a triumph for the forces, at home and abroad, of nativism, authoritarianism, misogyny, and racism ... [A] sickening event in the history of the United States and liberal democracy.'[22] He might have been describing the 9/11 terror attacks on America rather than a disappointing election result. Liberal film-maker Jim Jarmusch expanded further on that theme, explaining that 'the election of Trump is not only a tragedy for the United States. It is a tragedy for the world'.[23]

Meanwhile on American college campuses, students held a 'cry-in' (Cornell) or staged a collective 'primal scream' (Yale) to demonstrate their trauma and pain at the 'sickening' election of Trump. In turn, college authorities cancelled exams and offered their students counselling and time off to 'grieve', as if they were all the victims of an unexpected natural disaster, or perhaps an unheralded alien invasion.[24]

These reactions to both Brexit and Trump appeared different from the normal responses to an electoral setback. It was not simply that the losing side did not agree with the voters' verdict; it did not understand how they could possibly have reached it. The defeated establishment figures found the results not just uncomfortable, but entirely incomprehensible.

In short these seemed like more than ordinary electoral defeats. They signalled deep divisions and, above all, a cultural revolt – the near-total rejection of the values of the ruling elites by a sizeable section of the electorate. The subsequent response has been not to doubt the efficacy of those top-down values, but to question the wisdom of allowing the revolting masses to pass judgement on them from below.

## Two nations

The divides laid bare by the EU referendum in the UK and Trump's election in the US brought to mind the leading Victorian Benjamin Disraeli, later to become a Tory prime minister, who described in his novel *Sybil, Or the Two Nations* (1845) a state of 'Two nations between whom there is no inter-course and no sympathy; who are as ignorant of each other's habits, thoughts, and feelings, as if they were dwellers in differ-ent zones, or inhabitants of different planets. The rich and the poor.'[25] The divide today, however, is not quite such a black-and-white – or 'binary' – split caused simply by differences in wealth.

In the UK the divisions revealed by the EU referendum have been endlessly analysed along demographic lines, to show that young people were more likely to vote Remain than older people, or that higher votes for Remain were often found in areas with higher numbers of graduates from higher education and vice versa, or that most poorer people voted to Leave.

There is something in these attempts to analyse the divide. Class divisions certainly played an important part. But the focus on demographic divisions tends to make them appear permanent and immovable. The most important divides revealed by the results in the UK and the US, however, were surely the political and cultural splits across society today. This points up the importance of democratic debate – a clash between differing sets of values – to decide which direction our societies want to take.

Such meaningful debates have been scarce in recent times. Instead politics and public life in the UK, the US and other Western societies have increasingly become the preserve of a professional elite of officials, opinion formers and experts. This professionalised political elite relates to the rest of society through the media, if at all. Meanwhile millions of those patronised as 'ordinary people' have been treated as Others, deemed outside of politics and beyond the pale, their concerns marginalised and ignored.

If there is a gap between those who did and did not go to university in the UK, for example, it is not simply that Remainers are smart and Leavers 'too thick to vote'. It is more that those who participate in higher education – now around 40 per cent of young people in the UK – tend to be imbued with very different values, which reject most traditional ideas still dear to many in the world outside the university campus.

The new class of intellectual and moral elitists has been well described by the US writer Joel Kotkin as a 'Clerisy', a term he borrows from the English philosopher-poet Samuel Taylor Coleridge. Some 180 years ago, notes Kotkin, Coleridge described approvingly an educated, enlightened middle class that would serve a priestly function for society. He called them a Clerisy, adapted from *Klerisei*, a German word for clergy. One dictionary suggests that 'Coleridge may have equated *clerisy* with an old sense of *clergy* meaning "learning" or "knowledge"', which by his time was used in the proverb 'an ounce of mother wit is worth a pound of clergy'.[26] The poet wanted to reassert the authority of the enlightened elite he called the Clerisy over the base 'mother wit' of the masses.

Now, says Kotkin, what we have in both the US and Europe is a New Clerisy of middle-class professionals who dominate politics, culture, education and the media, 'serving as the key organs of enforced conformity, distilling truth for the masses, seeking to regulate speech and indoctrinate youth'.[27] Kotkin observed in the run-up to the 2012 US presidential election that: 'Many of [the Clerisy's] leading lights appear openly hostile to democracy ... They believe that power should rest not with the will of the common man or that of the plutocrats, but with credentialed "experts" whether operating in Washington, Brussels or the United Nations.' That hostility to democracy has only intensified over the past few years.

The Brexit vote marked a breakthrough revolt of 'the common man' and woman against the 'enforced conformity' preached by the New Clerisy. That it came as such a shock to the Clerisy was a sign of how little contact they had with the real world occupied by Other People.

They might have done well to note the report by David Cowling, former head of the BBC's political research unit, which was leaked just before the referendum. He noted that: 'There are many millions of people in the UK who do not enthuse about diversity and do not embrace metropolitan values yet do not consider themselves lesser human beings for all that. Until their values and opinions are acknowledged and respected, rather than ignored and despised, our present discord will persist.'

Cowling observed that 'these discontents run very wide and very deep and the metropolitan political class, confronted by them, seems completely bewildered and at a loss about how to respond ("who are these ghastly people and where do they come from?" doesn't really hack it).'

His report concluded that the EU referendum had 'witnessed the cashing in of some very bitter bankable grudges' but that throughout the campaign 'Europe has been the shadow not the substance.' The 'ghastly people' had simply seized upon the EU referendum and voted Leave as a way to express their long-held wide and deep discontent with the elite who so obviously despised them.[28]

A few months later, the November 2016 US presidential election marked another remarkable revolt against the New Clerisy's values of 'enforced conformism'. As with Brexit, the elitist view of Trump's victory as 'incomprehensible' only demonstrated how detached the US establishment had become from the lives and concerns of millions of Americans.

After the election, everybody suddenly started asking 'How could They vote for HIM?' It should not have been too difficult to get sensible answers beforehand. It was just that nobody had bothered to ask 'them'. Belatedly, some major media outlets did attempt the basic journalistic job of talking to voters. When the *Washington Post* asked its readers to give a brief post-election explanation of 'Why I Voted for Trump', it had soon received more than 1600 revealing responses.

Many of them were at pains to emphasise that, in the words of one voter, 'I do not 100 per cent love Donald Trump', and to disassociate themselves from his comments about women and wild outbursts about immigrants. They had voted not so much for Trump as against the establishment that ignored them and backed Clinton; his reported misogynistic remarks had not swayed them, for example, because they never thought or cared about him being a feminist anyway.

As forty-seven-year-old Nicole Citro of Burlington, Virginia, wrote in her contribution to the *Post*, she 'saw how the media, the establishment and celebrities tried to derail him' and hoped 'that I would be able to witness their collective heads explode when he was successful. Tuesday night [election day] was beyond satisfying to watch unfold.'

Elsewhere in the *Post*, sixty-one-year-old Diane Maus of Suffern, New York, expressed her anger at how the media discussion had given the impression that 'voting was a mere formality. The commentary was all about how Hillary Clinton was set to get down to business once the pesky election was over.' For Diane and millions of Trump voters like her, 'My vote was my only way to say: I am here and I count.'[29]

However, 'the media, the establishment and celebrities' still were not listening, or at least could not comprehend what was being said. The isolation of these types from the people they look down upon was well summed up by those last-gasp celebrity rallies for Clinton. They seemed seriously to believe that the image of Madonna singing a bad acoustic version of John Lennon's 'Imagine', interspersed with screeching 'No way, motherf*cker!' in Trump's direction, would make a difference on voting day. Imagine that …[30] Some appalled American celebrities swore to emigrate after Trump's election. But they already appeared to be living on a different planet from the people who had voted Trump in order to make the point that 'I am here and I count' in a democracy, just as much as Madonna or Beyoncé.

The votes for Brexit and Trump represented a revolt of the Others, a demonstration by the deplorables, against the Clerisy. At a loss to understand what those Others were talking about, the elites instead sought to impose their low opinion of voters as a judgement on the adverse voting results. Why had Remain lost the referendum and Clinton failed to become president? Clearly they could not accept that it was the fault of their unpopular politics, or that they had simply lost the argument. So the populace must be to blame, for losing its senses.

The problem became the revolting people, the *demos*. In which case the democratic system that gave them the chance to dictate to their betters must ultimately be at fault.

There have been differences in the masses-bashing responses to Brexit and Trump. But three common themes stand out. All are attempts to delegitimise the results and the voters who produced them.

The first theme is that the votes were a result of ignorance and disinformation in the age of 'post-truth politics'. The second is that the voters must have been motivated by bigotry, racism and hatred. And the third is that, given the above, allowing the votes of the *demos* to determine important issues is a threat to … democracy.

Let's look at these excuses in turn.

### 'Post-truth' politics for 'unqualified simpletons'

It has been widely argued and accepted that those voting for Brexit in the UK or Trump in the US must have been uninformed, 'low-information' people, emotionally gullible and easy prey to the lies of demagogues – now renamed 'post-truth politics'. As leading Labour politician Chuka Umunna summed it up, 'Both Donald Trump and the Vote Leave camp epitomised "post-truth politics"'.[31] This notion updates the prejudice expressed by ancient Greek philosophers that democracy entrusts too much influence to the ignorant, over-emotional and easily misled many at the expense of the wise and enlightened few.

Showing contempt for the masses is no longer the preserve of Roman generals and authoritarian governments. One striking feature of the resurgence of anti-democratic prejudices has been the leading role of liberal intellectuals. The more high-minded the commentator, it appears, the lower view they take of the masses and their apparently mindless antics in the voting booth. As elsewhere, the reaction to the UK referendum result set the pattern.

British intellectuals were in the vanguard of the anti-Brexit backlash. There was Professor Richard Dawkins, the leading evolutionary biologist, professional atheist, humanist scientist and scourge of blind-faith religionists everywhere. In the left-wing *New Statesman* magazine soon after the referendum, Dawkins the great humanist seemed unable to suppress his true feelings about that large slice of humanity who voted Leave as 'stupid, ignorant people'. He protested that 'it is unfair to thrust on to unqualified simpletons the responsibility to take historic decisions of great complexity and sophistication'. Presumably such decisions would be better left to complex and sophisticated minds such as the Professor's own.[32] The great atheist appears to think that the rest of the electorate should have blind faith in the wisdom of the expert priesthood.

Dawkins also protested (retrospectively of course) that 'the bar should be set higher than 50%' in referendums, as a way of diminishing the scope for democratic decision-making by unqualified simpletons: 'A two-thirds majority, or at least a threshold that lies outside the statistical margin of error, is one way to guard against this.' In other words, a minority should have a veto. It was left to psychology professor David Shanks to point out in a letter to the *Statesman* that Dawkins himself was 'guilty of a statistical error'; margins of error have to do with samples in opinion polls, not actual votes: 'The concept of a margin of error has no meaning when an entire population expresses its opinion.'[33] Dawkins's 'statistical error' looked like a classic example of an eminent scientist using scientific-sounding language to justify his personal opinion about a political issue on which he has no more claim to expertise than any other voter.

Nobody seemed more agitated about the Brexit vote than the normally unflappable 'leading man of the Left', philosophy Professor A. C. Grayling, who wrote to every Member of Parliament (apparently in the name of his students), demanding that they take a vote to ignore the result and remain in the European Union. In his 2009 book, *Liberty in the Age of Terror*, Professor Grayling had warned of the need to defend our hard-won democracy, rights and 'Enlightenment values' against the encroachments of the security state.[34] Now, by contrast, he called upon the authorities to usurp the referendum result and secure Britain's membership of the EU against the encroachments of the unenlightened people.

Writing in the *New European*, house journal of the Remainers, where he was heralded as 'Britain's leading philosopher' (surely that should be 'Europe's'?), Professor Grayling laid into the 'uninformed, hasty, emotional and populist ways' Leave had won, based on mere 'demagoguery and sentiment'. The good Professor's repeated attacks on the 'emotional' attitudes of the other side might seem ironic, since nobody wrote more emotionally about it than him. Presumably the majority of those who voted had simply expressed the incorrect emotions.[35]

The real problem, according to Professor Grayling, is that 'the majority of people are "System One" or "quick" thinkers' who 'make decisions on impulse, feeling, emotion, and first impressions'. This left them open to 'manipulation' by demagogues peddling 'post-truth politics' and 'downright lies', who had persuaded them to support the 'lunatic' notion of Brexit. What we need, apparently, is to pay more heed to 'System Two' or 'slow' thinkers, 'who seek information, analyse it, and weigh arguments in order to come to decisions' – such as voting Remain, of course. It seems that 'System Two' voters are naturally more equal than others.[36]

Would the professor prefer to see the re-introduction of special university seats in the UK parliament, which gave graduates of Oxbridge and other top universities an extra vote until they were abolished by the ghastly Labour government after the Second World War?

The emphasis of many critics was on the 'Brexit lies' of the Leave campaign and how they had led gullible voters astray. This was apparently proof that we live in the age of 'post-truth politics'. Indeed not long after the referendum and the election of Trump, Oxford Dictionaries announced that 'post-truth' was its international word of the year for 2016. The *Oxford English Dictionary* defines this award-winning expression to mean 'relating to or denoting circumstances in which objective facts are less influential in shaping public opinion than appeals to emotion and personal belief'. The Remainers reduced this to the basic claim that their 'objective facts' and truths had lost out to Leave's 'appeals to emotion' and outright Brexit lies.

In the aftermath of the June vote there was much dark talk about the need for the enlightened to tackle 'post-truth' politics. The UK Electoral Reform Society produced a damning report on the referendum campaign, claiming that there had been 'glaring deficiencies' in the facts offered by both sides which had left voters 'feeling totally ill-informed'. The ERS report concluded with the Orwellian-sounding proposal for an 'official body ... empowered to intervene when overtly misleading information is disseminated' in future political campaigns, presumably to protect gullible voters from their own ignorance by force-feeding them official facts. Perhaps it should be called the Ministry of Truth?[37]

What's the truth about those 'Brexit lies'? There were of course exaggerated claims and flights of fancy on both sides of the EU referendum: from the official Leave campaign's fantasy of a quick extra £350 million a week for the NHS, to the Remain campaign's horror stories of imminent economic depression; from Boris Johnson's comparison of the EU with Hitler, to David Cameron's warning that a vote for Brexit would delight ISIS and could start the Third World War.

Much of this is the overblown-but-normal cut-and-thrust of heated political debate in an electoral firefight. Voters do not need to be protected from such stuff by the wise men and women of the European Commission, the ERS or any other fact-checkers or 'official body' set up to decide The Truth on our behalf. What voters need is to be left alone to listen to all the arguments, join in the debate as they see fit, and ultimately decide for themselves what they consider to be truly in their own, and their society's, best interests. In this sense, the EU referendum looks like an advert for the virtues of popular democracy.

Indeed, far from being duped by Brexit lies, the Electoral Reform Society report on the campaign revealed that most voters they spoke to had a 'highly negative' view of the official campaigns, and said the top politicians' appeals had made 'no difference' to how they voted in the end. Where there had been any effect, it was most often the opposite of what the politicians intended; the interventions by top Remain-backing figures, from Cameron and Corbyn to Nicola Sturgeon and Barack Obama, had all made people marginally more likely to vote Leave.

They should have known. A pre-referendum poll conducted by Ipsos MORI was already revealing about the likely impact of experts and political elites. In May 2016 they asked respondents to answer the question: 'Who do you trust on issues related to the referendum on EU membership?'. The winner with 73 per cent approval was 'Friends and immediate family'. Other strong runners included 'Work colleagues' and notably 'The ordinary man/woman in the street', both with 46 per cent approval ratings. Lower down came 'Leaders of large business' (36 per cent) and 'Civil servants' (29 per cent). Rooted in the relegation zone of this public trust table were 'Journalists' on just 16 per cent and lastly 'Politicians generally' with a miserable 12 per cent – in a much lower league than those 'ordinary' men and women in the street.[38]

(The one odd note in this expert-bashing survey of public trust was that 'Academics' came second behind 'Friends and immediate family', with 66 per cent, showing that these experts are still held in relatively high regard. Not high enough, mind you, for the UK's overwhelmingly pro-Remain academic community to make a difference to the ultimate referendum result.)

What, then, was 'the truth' that the Remain campaign had tried and failed to sell to voters? Essentially they sought to displace any discussion of the wider political issues of democracy and sovereignty, and focus the debate on their dire predictions of economic doom if the UK voted to leave the EU, in a bid to bully supposedly simple-minded voters into obedience. The message echoed the fatalistic view that was captured by Margaret Thatcher in the 1980s and has effectively been repeated by every UK prime minister since; that 'There Is No Alternative' to the economic status quo, so forget about choice, lie back and think of the European Single Market.

The possible economic consequences of Brexit remain unclear, and were certainly uncertain in advance of the referendum. So what were the fear-mongers' warnings of economic catastrophe really saying to UK voters? That it does not matter what you think or want, the global financial markets must decide. Share prices and the exchange rate of the pound are the determining factors of history. Your vote is worthless by comparison; swallow your medicine and watch the markets.

Yet for all this, what was remarkable was that the majority of the 72.2 per cent who voted declined to be swayed or bullied into submission. They kept their eyes on the bigger issues of sovereignty and democracy and voted Leave because they wanted more control over their own lives, UK politics and the country's borders. It was not about the electorate's ignorance or economic illiteracy. Millions made the entirely rational calculation that these reasons were important enough to support Leave, even if the immediate economic impact was uncertain and might prove adverse. Contrary to what the doom-mongers claim about 'post-truth' politics, it is perfectly reasonable to decide that the possibility of a fall in the value of the pound could be a price worth paying for an increase in democracy and sovereignty. Just as it was perfectly rational for others to vote Remain because they judged it to be in their best material interests.

Even before the US presidential election, many American critics were already following the lead of the Remain campaign and complaining about the influence of 'low-information' (code for low-intelligence) voters and the emotive 'post-truth politics' allegedly being practised by the Trump campaign. A week before polling day, academic Marci A. Hamilton caught the mood of exasperation when she asked in *Newsweek*, 'Why are white, uneducated voters willing to vote for Trump?' Answering her own question, as most academics like to do, she concluded, 'I would posit that it is also because they have not been adequately educated to understand just how dangerous a President Trump would be to the Constitution.' In other words, they had not been 'adequately educated' by the likes of Hamilton to swallow whatever they were now being told by the same people.[39]

In the shocked reaction to Trump's election, these latent prejudices about the influence of 'low-information' American voters came pouring forth. Author and radio personality Garrison Keillor snorted that 'Trump has won. Let the uneducated have their day'.[40] For Georgetown professor Jason Brennan, who bluntly blamed Trump's victory on 'low-information white people', the election placed a big question over democracy itself: 'Democracy is supposed to enact the will of the people. But what if the people have no clue what they're doing?'[41]

The arguments about 'low-information voters' and 'post-truth politics' provided a convenient excuse for the elites' failure to get enough voters to do their bidding. After all, what hope have you got of convincing people if they are just too stupid and uneducated to recognise what is both true and truly in their interests?

Unfortunately, observed Republican commentator Rob Schwarzwalder, even in its own terms the argument that Trump won because of uneducated voters 'has the disadvantage of being untrue'. In the 2016 election, the Pew Research Center's exit polls found, college graduates did favour Hillary Clinton over Donald Trump by 52 to 43 per cent (though white graduates were for Trump by 49 to 45 per cent), while 52 per cent of voters without a degree voted for Trump with 44 per cent for Clinton. Yet in the previous presidential election in 2012, graduates voted for Democratic president Barack Obama over Republican challenger Mitt Romney by 50 to 48 per cent, while those without a college degree favoured Obama by a wider margin, 51 to 47 per cent. It is hard to recall many experts denouncing Obama's win and blaming it on these 'uneducated' voters.

But then this discussion is not really about the statistics of college degrees and votes. It is about the elites recycling age-old prejudices about the dangers of allowing the ignorant, emotional masses to exercise control, in order to excuse their own failings as somehow being a serious flaw in democracy.

This condescending attitude towards the mass of people goes some way to explaining why those voters – who are quite intelligent enough to know when they are being patronised and insulted – refused to do as they were told at the polls. As conservative commentator Fred Weinberg wrote, the media's basic message to Trump voters was: 'You're Uneducated and Deplorable'. Since most media people 'never talk to real people', they didn't get the resentment felt by millions of Americans at 'being told we live in "flyover country" … comprised of "uneducated" white males who do not understand that we need to be told how to live by "journalists" who live in the progressive bubble. Or by their elected friends.'[42]

## Playing the new race card

The second widespread attempt to explain away the 'disaster' of the referendum result and the 'tragedy' of the US election has also focused on the shortcomings of the electorate. People who voted for Brexit and for Trump, we are blithely assured, must have been racists, xenophobes and Islamophobics. In which case their votes should be seen as morally illegitimate at least, if not legally suspect.

The pattern was set in the run-up to the EU referendum. Reports that some England football fans involved in trouble during the European Championships in France had been heard chanting 'F*ck off Europe, we're all voting Out!' were seized upon as evidence that Leave supporters were basically an ignorant mob of hooligans, xenophobic and brutish, only a couple of pints away from launching a racist pogrom. The small-minded prejudices displayed like a football flag here were those of leading Remainers towards beer-drinking, football-watching working-class voters, who appear to them far more alien than suave Brussels bureaucrats.

A week before the referendum, pro-Remain Labour MP Jo Cox was murdered in the street by an apparently mentally disturbed man with a collection of Nazi books and paraphernalia, shouting about 'traitors' and 'Britain first'. The killer was quickly branded a 'Brexit nutter' and pointed to as proof of the hatred and bigotry allegedly endorsed by the Leave campaign. The violent crimes of one racist madman thus became twisted into evidence against millions of sane, non-violent voters.

Almost immediately after the referendum result, a new scare started over a reported spree of 'hate crimes' against immigrants in various parts of the UK.

The political elite seized upon the allegations of racism with relish, to try to prove that some bigoted voters really should be seen as less equal than others. 'I'm afraid it has to be said that there has been a vote from white working class Labour supporters. They have voted in the face of the fact that they have probably never even seen a migrant and it's the fault of politicians,' said Tory MP Anna Soubry. Leading Remainer Soubry, who was then minister for small businesses (not small minds?), told the BBC that the Brexit campaign had 'unleashed' a latent wave of anti-immigrant racism.[43]

Labour shadow health secretary (soon to become shadow home secretary) Diane Abbott responded to these attacks on her party's traditional white working-class voters – by not only agreeing with the critics but going further. Abbott told a 'Brexit: Unite Against Racism and Hatred' event at the Labour Party conference that Labour MPs should not even discuss the issue of immigration with Leave supporters. Such people would not be satisfied because 'what they really want is to see less foreign-looking people on their streets'.[44]

This cross-party political consensus against the white working classes seemed unsure whether they had voted Leave because they saw too many immigrants on their streets, or in spite of having never seen any. But both Tory and Labour Remainers apparently agreed that the Brexit vote had been a demonstration of British racism and bigotry.

Does anybody seriously believe that 17.4 million UK voters backed Leave for racist motives? If not, how many million racists do they think there really were among Brexit voters? The only thing running wild here was not a racist mob but the dark imaginations of political elitists.

The belief that voting to Remain was an anti-racist decision while Leavers must have been anti-immigrant reveals more about the one-eyed view of the anti-Brexit lobby. What do they imagine is so staunchly pro-migrant about the EU? If the European Union is such an open-borders institution as its officials insist, why are so many migrants barred from entering it drowning in the Mediterranean Sea?

Immigration was an important factor for many Leave voters, though hardly the obsession it has been made out to be; a post-referendum ComRes poll found that 34 per cent said immigration was their main concern, with 53 per cent instead prioritising the 'ability of Britain to make its own laws'.[45] Most of those concerned about immigration, however, did not see the issue in the crudely racist, send-'em-back style of the 1970s. In August 2016 a think-tank poll found a remarkable 84 per cent of British voters wanted EU migrants living and working in the UK to be allowed to stay after Brexit – including 77 per cent of Leave voters.[46]

The truth is that Britain in 2016 was a far more tolerant and anti-racist society than at any time in its history. Problems of overt racist abuse and violence bear no comparison to the bad old days of the 1970s, when I grew up in a suburban Surrey where racism was not so much acceptable as obligatory, and the 1980s, when some of us on the Left in politics organised to help defend immigrant families under threat of being burned out of London housing estates.

Despite all the warnings of racism on the rise, every serious survey of attitudes to race and ethnicity in British society tells the same story of growing tolerance today. One article published in October 2016 summarised various findings: only one in ten Brits now 'endorse nakedly racist views'; the proportion of the English public 'most hostile to immigration' for racist reasons has shrunk from 13 per cent to 7 per cent; while the World Values Survey now 'rates Britain as one of the most racially tolerant countries in the world'. None of which prevented the Remainer newspaper in question publishing the article under a headline which declared, contrary to all its own evidence, that post-referendum 'Britain is becoming mean and small-minded'.[47]

The political and media panic about an alleged wave of 'hate crimes' after the referendum appeared equally dubious. A small handful of serious attacks, which may or may not have had anything to do with the referendum result, were mixed in with reports of many other minor or questionable incidents to create the impression of a brewing pogrom, with some commentators even indulging in horror fantasies about 'the rise of fascism across the country'.

In October 2016, the Home Office reported a 'sharp increase' in hate crime after the referendum; there had been 5,468 hate crimes in July that year, a shocking 41 per cent up on the figures for July 2015. What were these crimes? They were alleged incidents reported to the police, often through phone, email or social media hotlines. They had not been investigated, far less tried as crimes in a court of law.

Instead the police simply record everything they are told about as a hate crime, without any need to question or investigate at all. The Operational Guidance for police forces explains: 'For recording purposes, the perception of the victim, or any other person, is the defining factor in determining whether an incident is a hate incident … The victim does not have to justify or provide evidence of their belief, and police officers or staff should not directly challenge this perception. Evidence of hostility is not required for an incident or crime to be recorded as a hate crime or hate incident.'[48]

So, unlike other crimes, if anybody at all says anything at all is a hate crime, the police must record it as one, no 'evidence of hostility' required or questions asked. What was that about 'post-truth' politics and the downplaying of 'objective facts'? Indeed, given this subjective system, and the way that the police and the mayor of London made high-profile appeals to report any suspected hate crimes after the referendum, the wonder might be that the statistics for reported incidents showed an increase of only 41 per cent.

This looked like the twenty-first-century equivalent of the mugging panics of the 1980s. Then, every black inner-city youth had been looked at in fear as a potential mugger. Now every suburban white working-class youth was being viewed with similar dread as a potential hate criminal. That one prejudice was an expression of racism and the other ostensibly of anti-racism does not alter the fact that both are expressions of bigotry. In effect the scaremongers were playing a new version of the race card.

In America, protesters angry at the election of Donald Trump lost no time in branding his millions of voters as racists, 'white supremacists' and even Nazi sympathisers, and therefore unfit to choose a president.

Alongside the allegation that a vote for Trump was a 'hate crime', post-election protesters chanted 'No Trump, No KKK, No Fascists USA!'. Meanwhile on CNN's election night coverage, commentator Van Jones made international headlines and inspired many imitators by immediately branding the Trump vote as a racist 'whitelash' after eight years of a black president.[49] Nobel Prize-winning economist Paul Krugman tweeted his shock at discovering the extent of the 'deep hatred in a large segment of the population'.[50] Jamelle Bouie, chief political correspondent for *Slate* magazine, rejected the very suggestion that any Trump voters might be viewed as non-racist, good people: those who once 'brought their families to gawk and smile' at racist lynchings, he wrote, were 'the very model of decent, law-abiding Americana. Hate and racism have always been the province of "good people"'.[51]

As with the dismissal of 'low-information voters', much of this stuff says more about the prejudices of the elites than about political realities. True, the same authoritative exit polls suggest that non-Hispanic white voters backed Trump over Clinton by 58 to 37 per cent. But the headline-grabbing argument about this being due to a racist 'whitelash' lasts about as long as it takes to glance at the figures. Both of the main 2016 presidential candidates were white; Hillary Clinton got a lower percentage of white votes than Barack Obama did in 2012 – and Trump got a lower percentage of white votes than Mitt Romney did that year; indeed, as one rational anti-Trump blogger put it, 'The only major racial group where he didn't get a gain of greater than five per cent was white people.'[52]

As for the idea that anybody voting for Trump must have been a racist, a white supremacist, an 'alt-right' zealot or a Ku Klux Klan fan, and that the ghost of Hitler was now stalking the USA – these ridiculous claims make some of the Donald's own ramblings and rants seem almost reasonable by comparison. (Note to the historically confused: whatever Hitler was – a genocidal Nazi racial supremacist – he could not be accused of being a buffoonish celebrity loudmouth who made up and tore up policies as he went along.)

Take the alleged Ku Klux Klan links, which many anti-Trump protesters seemed keen to highlight. According to Wikipedia, 'As of 2016, the Anti-Defamation League puts total Klan membership nationwide at around 3,000, while the Southern Poverty Law Center puts it at 6,000 members total.' Trump, remember, won more than 60 million votes. Which means the top estimate of KKK membership in the entire US is equivalent to less than 0.01 per cent of his support.

No doubt any group of 60 million people will include racists and all manner of others of dubious outlook. However, anybody not blinded by their own prejudices would have to recognise that, as with the UK, attitudes towards race across American society have changed fundamentally in recent decades. In 1960, for example, some 50 per cent of white Americans told Gallup pollsters that they supported racial segregation in schools and would move home if a black family moved in next door; by the late 1990s that figure was down to between 1 and 2 per cent.

As even the film-maker and Democratic Party propagandist Michael Moore was moved to concede, it makes no sense to equate voting Trump with racism: 'You have to accept that millions of people who voted for Barack Obama – some of them once, some of them twice – changed their minds this time. They're not racists. They twice voted for a man whose middle name is Hussein. That's the America we live in.'[53]

Indeed it is – an America ruled over by a Democratic Party administration for the previous eight years. Might that political experience have had something to do with why many Americans were prepared to vote for somebody like Trump this time? Could his election be seen as the legacy of Obama's and Clinton's politics in power? No, that was unthinkable. Far simpler to blame it on a racist 'whitelash' against Obama's skin colour.

As a writer on 'cool' online magazine *Vox* declared, 'Trump's win is a reminder of the incredible, unbeatable power of racism.'[54] How convenient. If the popular power of racism is really deemed 'unbeatable', then Clinton and the Democrats surely cannot be blamed for the incredible, catastrophic failure of their campaign.

The dangerous driving force in this discussion is not race hate but fear and loathing of the masses. Branding opposing views as offensively racist or supremacist has become a trendy all-purpose insult, a way to delegitimise votes and opinions that are not to your taste. They can then simply be dismissed, or possibly banned, rather than debated; after all, who wants to engage with racist nutters or Klan fans?

Such permissive use of the insult 'racist!' trivialises the term, and diminishes the importance of real racism. It can sound like a simple reversal of old-fashioned racist notions about anything black being inherently inferior.

After the election Gretchen Reiter, who describes herself as a professional 'Washington insider' living in rural America, wrote of her anger during the campaign at seeing 'my friends and family reduced to a label given by elitist, intolerant talking heads: uneducated white people'. For her, 'the last straw' came a few days before the election, when she heard a *New York Times* columnist on PBS say 'that voters are supporting Trump because of their "gene pool". It was insulting and ignorant'.[55]

As insulting and ignorant, some might think, as old-school racial notions about the 'natural' inferiority of those enslaved and excluded from power.

### 'Democracy' against the *demos*

The third common strand in the backlash against the Brexit and Trump votes is the attempt to justify these attacks on the *demos* – the people – in the name of democracy.

Few in the West feel able explicitly to reject democracy these days. So the trend is to try to redefine its meaning instead. According to this new definition, 'upholding democracy' means protecting the political status quo – if necessary, against the people, in whose name democracy exists.

Professor A. C. Grayling was once more to the fore here, loftily informing British MPs that it was their 'democratic duty and responsibility to reaffirm continuation of the UK's EU membership'. To do otherwise would apparently 'subvert our representative democracy and our constitution'.

How, exactly, could it 'subvert our representative democracy' to accept the democratically expressed will of 17.4 million people? The key word here is 'representative'. As we shall see, our society's idea of democracy has been redefined over the years to mean a system where a political elite can 'represent' the people as it sees fit.

Grayling captured the elitist essence of this order by arguing that the EU referendum was only 'advisory', that parliament is sovereign, and that most MPs disagreed with their constituents and backed Remain. In which case, he insisted, the constitution decrees that the people's representatives should ignore the result and vote to remain in the EU, since British democracy was actually about power being in the hands of 645 MPs and the nearly 800 peers in the House of Lords, not the voters.

To accept the referendum result, the wise professor suggested, would be to give in not to democracy but to 'ochlocracy'; a word meaning 'government by the populace' or, in elite-speak, 'mob rule', which has barely been used since the oligarchs – the powerful few – of ancient Athens looked down their aristocratic noses in horror at mass democracy.

Grayling was clear that mere numbers are not the decisive issue in a democracy, since a lack of proper education apparently prevents the general public from expressing 'the general will', which he seems to think a few people like him are better placed to understand. Oddly, however, the Professor also insisted that sixteen-year-olds should have been given the vote in the referendum; no doubt because he thinks teenagers are better informed, educated and wiser than their elders, and not simply on the assumption that they might have been more likely to vote Remain.

Similar 'democratic' arguments for overthrowing the referendum result were advanced by all the other voices arguing against the Brexit vote, many of whom appeared to believe that the granting of democratic rights means you keep people voting until they arrive at the right result.

The Brexit-bashers all seemed keen to draw parallels between the Leave campaign and the Trump crusade in the States. Yet it was their refusal to accept the referendum outcome that more closely chimed with the pre-election attitude of Donald 'I'll respect the result – if I win' Trump.

There were the handful of backers who funded a legal bid to get judges to rule that, regardless of the 'advisory' referendum result, the government could not trigger Brexit without the backing of MPs and lords in parliament. In true Newspeak-style, this stunt was called 'The People's Challenge'. Thus the real people's challenge to the technocratic political elite, as seen in the referendum result, was threatened by a self-interested 'People's Challenge' and dressed up in the finery of the Royal Courts of Justice. In November 2016, three high court judges ruled in the legal claimants' favour, and declared that the government must have the approval of parliament before it could trigger Article 50 and begin the Brexit process. In effect the Lord Chief Justice Lord Thomas of Cwmgiedd, the Master of the Rolls Sir Terence Etherton and Lord Justice Sales took it upon themselves to shelve the democratic decision of the electorate. The people might have spoken, but three law lords had told them to shush and let their betters talk among themselves. This was a microcosm of the trend for unelected bodies to assume authority over the *demos*.

Then there was the apparently four-million-strong online petition calling on parliament to hold another referendum that would require a larger margin of victory (a petition initially launched before the referendum by a Leaver who feared a narrow Remain win, which suggests that faith in popular democracy is not necessarily stronger on the other side). When this demand was debated in parliament, as petitions which garner more than 100,000 signatures must be, Tim Farron MP, leader of the pro-EU Liberal Democrats, said (seemingly with a straight face), 'We demand that the British people should have their say on the final deal in a referendum!'[56] Never mind that the people had just had their say in a referendum; that could not be final, since too many people had misspoken and said 'Leave!' Not for the first time it might occur to some that the

Lib Dems have a strange idea of the meaning of liberal democracy.

In similar vein was the letter signed by around a thousand top lawyers, demanding that parliament must decide (i.e., vote for Remain). As the Queen's Counsel who organised this initiative, Philip Kolvin QC, explained, 'In times of crisis people often turn to lawyers to ask them how we should behave in society.'[57] Of course we do. The notion that the opinions of 1,000 lawyers on 'how we should behave in society' could outweigh those of 17.4 million voters mostly without law degrees summed up the some-are-more-equal-than-others essence of the backlash against Brexit. Even though this top legal advice was unusually offered free of charge, the price to pay for a free society accepting such guidance would be far too high.

In parliament, meanwhile, a cross-party alliance – including Labour MPs such as leadership contender Owen Smith and Tory lords such as Baroness Patience Wheatcroft – was busily conjuring up constitutional and 'democratic' arguments as to why they should act to ignore, overthrow or otherwise seek to reverse the referendum result.

Leading Labour MP David Lammy wrote that 'we cannot usher in rule by plebiscite which unleashes the "wisdom" of resentment and prejudice reminiscent of 1930s Europe'.[58] Note the inverted commas around the word *wisdom* when applied to the masses. For the likes of Lammy, it appeared, overturning the EU referendum result was now on a par with defending democracy against fascism.

This was a sure sign of what democracy has come to mean: a hollowed-out, narrowly defined system of rule which denies a meaningful say to the *demos* – the people – from whom the idea takes its name, and concentrates control in the hands of an elite that looks more like the privileged oligarchy of ancient Athens. Where the Greeks practised direct democracy, we have long been told that representative democracy is better suited to modern times. Now it seems we are left with an increasingly unrepresentative form of democracy – and when people revolt against the orders of the elite, the response is to try to make our democracy less representative yet.

Those arguing in the language of constitutional law to delay or reverse Brexit boasted impressive legal and academic credentials. Yet in the language of real democracy, their arguments against accepting the referendum result were bogus. They constituted a legalistic mask to disguise the authoritarian intent.

The referendum was in no way merely an 'advisory' measure, to be ignored or accepted at parliament's pleasure. It was legislated for by parliament in a 2015 Act, passed with overwhelming support, which made no mention of a referendum being advisory, and conveyed the clear understanding that the government would give effect to the result. If there was any doubt about that, during the referendum campaign the Conservative government which commanded a majority of MPs sent out a propaganda pamphlet to every household, clearly stating the government's belief (backed by the official opposition parties) that Britain should remain in the UK. This document concluded: 'This is your decision. The Government will implement what you decide.'[59] Nothing advisory or open to interpretation there.

The issue of parliamentary sovereignty became the major smokescreen for attacking democracy. The apparently lofty and learned argument is that the UK parliament is sovereign, so that the government could not act to implement Brexit without gaining the approval of MPs and Lords. Under Britain's largely unwritten constitution, however, ultimate sovereignty still rests with the sovereign of this constitutional monarchy, exercised through the device of the Crown-in-parliament. The Royal Prerogative gives the government the power to do all manner of things in the name of the Crown, from waging war to signing treaties, without parliamentary approval.

The royal power that the Royal Prerogative grants to a government should be a big problem for anybody seriously concerned about democracy in Britain. It is the main reason why some of us have always favoured abolishing the monarchy. Curiously, however, it had never seemed much of a problem before the Brexit vote to those arguing against accepting the referendum result. Leading Remainers such as ex-New Labour prime minister Tony Blair and his allies certainly never had a problem with using the Royal Prerogative to launch destructive foreign wars while in office. Yet now they suddenly object to its use to pursue Brexit.

Most strikingly, few of the new champions of parliamentary sovereignty appeared in the slightest bit bothered about having that sovereignty overridden via the EU, by both European bureaucrats and British governments, through the previous forty years. It seems that they only became excited about the need to defend parliamentary sovereignty against the people.

These pseudo-democratic arguments were just a device to discredit the referendum which had been a genuine exercise in mass democracy. By August 2016, Labour front-bench MP Barry Gardiner could even accuse the new Tory prime minister Theresa May of acting 'with the arrogance of a Tudor monarch' by insisting that she could implement a form of Brexit without a further vote of MPs.[60] Leaving aside for a moment the fact that Mrs May was a Remain campaigner who did not want

Brexit; Mr Gardiner may have studied some different history from me, but I do not recall the well-known Tudor monarch Henry VIII reluctantly breaking with the Church of Rome or beheading two wives and assorted enemies because the people demanded it in a referendum.

These issues highlight bigger problems with British democracy, on both sides of the Brexit debate. For one side, it appears, representative democracy means that MPs should have the right to do as they see fit, regardless of the referendum result, in our increasingly unrepresentative system. For the other side, it seems, the Royal Prerogative gives the government the right and power to do as it pleases – including either implementing or delaying Brexit – not in the name of the people, but of the Crown.

Neither side measures up to the standard of meaningful democratic politics. But however we see these problems, the solution to the 'democratic deficit' in the UK cannot be even less democracy. That is what it would mean if we were to allow the elites to undermine or ignore the clearly expressed will of the majority who voted in the EU referendum. (If politicians now claim that 52 per cent of those who voted is not a legitimate mandate, by the way, then the UK has not had a legitimate government in living memory, since no party since the Second World War has ever achieved as many as 50 per cent of the votes cast.)

We need to find new ways to bring British democracy to life and make it mean more. Instead we are faced with a situation where democracy means so little that the Left can join with Tories in looking to the House of Lords to thwart the popular Brexit vote.

The unelected, unaccountable character of the upper chamber ought to be a problem for anybody who believes in democracy, making the Lords prime candidates to be voted into the dustbin of history. Yet that, it seems, is precisely why the unelected peers are considered so well qualified to 'defend democracy' against the referendum mob! As Baroness Wheatcroft spelled it out, the House of Lords was better placed to lead a 'rebellion' against Brexit because it is unrepresentative and unelected. (The Conservatives, despite winning a majority in the House of Commons in the 2015 general election, had only a minority of 254 peers out of an inflated total of 798 in the Lords, while the openly pro-EU Liberal Democrats, then reduced to a rump of just seven elected MPs, could still boast 105 unelected members swanning about in the House of Lords.)

Baroness Wheatcroft gave the game away when she boasted that, 'with no constituents to fear', the Lords would be freer than the Commons to vote against the wishes of the electorate. It is fear of the mass of constituents that drives anti-democrats of every political stripe to seek refuge in the Lords, while claiming to be upholding parliamentary democracy.[61]

After the American election, the reaction against Trump voters also adopted the bogus language of democracy to disguise its anti-democratic intent.

Almost as soon as the overall result of the 8 November elections became clear, the cry went up that Trump would not be a legitimate occupant of the White House. Film-maker Michael Moore spoke for many top Democrats when he denounced the Donald as 'an illegitimate president' who 'does not have the vote of the people'.[62]

Anti-Trump protesters angrily pointed out that, while the Republican candidate had won a majority in the electoral college – the system the US uses to elect its president indirectly – Democratic Party candidate Clinton had won a larger share of the popular vote. By the end of November, with late votes still being counted, Clinton had some two million more votes than Trump – about 2 per cent of the total – but the way these votes were distributed between states meant the Republican had easily carried the electoral college by 306 votes to 232.

Within days of the election, a *Washington Post*-ABC poll found that one in three Democrat voters believed Trump's win was 'illegitimate', with 27 per cent of them feeling 'strongly' about it.[63] Those feelings appeared strongest within the metropolitan strongholds of the Democratic elite, where both their votes and media-focused protests against the result were concentrated.

Those modern tools of passive political activism, the online petitions, quickly began gathering support, calling on the 538 members of the electoral college to go rogue – or act as 'faithless electors' – and refuse to endorse president-elect Trump when they congregated on 19 December, even if voters in their state had supported him. The largest petition of this sort on Change.org quickly gathered more than 4.5 million signatures, demanding that the electoral college make Hillary Clinton president because 'SHE WON THE POPULAR VOTE'. Meanwhile college electors reported being 'bombarded' with phone, email and social-media messages calling on them to ignore their electorates' wishes and vote against Trump. Some Democrat electors themselves admitted to lobbying for their counterparts in the college to vote for 'Anybody But Trump' and switch support to a more 'respectable' Republican such as Mitt Romney.

As with the anti-Brexit forces in the UK, the anti-Trump protesters were using the language of representative democracy in an instrumental way, to justify their attempt to overturn a result they did not like.

The US electoral college does indeed represent a strange, distorted and undemocratic brand of representative democracy. That is precisely why it was established in the first place – to provide a potential brake on popular sentiments of which the US elites do not approve.

The Founding Fathers who led the American revolution against British rule from 1776 and established the US as an independent republic were fearful of 'too much' popular democracy. They created a system of 'checks and balances' that could, if necessary, restrain the people in the name of representation. As James Madison wrote, the US system was founded to give the political elite powers to stymie the electorate when people were 'stimulated by some irregular passion' to 'call for measures which they themselves will afterwards be the most ready to lament and condemn'.

The constitutional checks and balances were put in place to check the power of the people and to counterbalance the will of the majority. The powerful Supreme Court is one arm of this system. Another is the Senate, the upper chamber of Congress which gives two seats to every US state, regardless of population size, and thus enables smaller, rural and generally more conservative states to outvote the big urban centres.

And another elitist arm is the electoral college, through which electors, nominated from each state on the basis of the election result, cast the final vote for the next president. This likewise favours conservative smaller states and also gives a few hundred electors – members of the political establishment appointed by the major parties – the potential to overturn the election result.

The electoral college has never done that, although there have now been five elections where the new president lost the popular vote. The most recent one pre-Trump was in 2000, when Republican George W. Bush became president despite winning fewer votes than Democratic challenger Al Gore. Some researchers claim there was also a sixth occasion, in 1960, when Republican Richard Nixon might have won more votes than iconic Democratic president John F. Kennedy, but the close-run result was left ambiguous by the confusing electoral system in the state of Alabama.

Despite the electoral college's evident democratic shortcomings, however, there have been few serious attempts to reform it. America's powerful elites still prefer a system with inbuilt brakes that can prevent popular sentiment running riot.

Yet there was little that was democratic about the sudden upsurge of protests about the US system after the 2016 election. Those happy enough with the electoral college when it delivers results they want were simply furious because it allowed for the election of the despised Donald. Indeed, their demand for 538 members of the electoral college to overturn the votes of 62 million 'ill-informed' Trump voters represents a new face of anti-democratic politics in the US and the West.

The sudden outburst against the 'illegitimate' election result was really no more democratic than the electoral college itself. Few of those protesting or signing petitions were concerned about the system until it enabled Trump to win. We might recall how, before election day, these were the people outraged by Trump's suggestion that he might not recognise the result if he did not win.

Yet overnight, leading Democrats would have us believe they were genuine democrats who just wanted to uphold the popular vote. How? By using the undemocratic electoral college to overturn the election result. Their loud talk of Trump being an 'illegitimate' president was really a coded attack on the millions of deplorable Americans who cast their 'illegitimate' votes for him, rather than being any principled defence of American democracy.

Bill de Blasio, the Democrat mayor of New York city, told CBS it was 'inconceivable' that Clinton had been denied the presidency despite winning more votes: 'It doesn't make sense. And it's supposed to be in our constitution: one person, one vote. That's not what happened here.'[64] Anybody might have imagined that this system had just been invented to get the Republican candidate elected, rather than being the same one under which every Democratic Party president has entered the White House. Everybody's vote has never been of equal value under a system designed to restrain democracy in the name of representation. These sudden converts to electoral reform were only objecting now because too many of those persons had cast their one vote for Trump.

The same seemed true of the high-level demands for recounts of votes in several key states, supposedly because the polls might have been 'hacked' by Russian cyber-terrorists. Can anybody imagine such 'principled' protests in defence of American democracy being backed by the establishment if Clinton had been elected?

Endorsing claims that Russia somehow hacked the US election 'to promote a Trump win', one liberal blogger announced that 'the only Constitutional solution available to us is for the electoral college to serve the function that the Framers intended for it, namely to serve as a check on elections gone wrong.'[65] Thus the radical wing of American liberalism lined up with the most conservative Founding Fathers in their determination to halt the 'wrong' election results. What ultimately unites them is contempt for millions of voters whom they cannot comprehend.

Of course, Trump is no champion of American democracy – commentators had fun unearthing his tweets calling for a 'revolution' to overthrow Obama's election win in 2012.[66] He is an illiberal at heart who poses a potential threat to precious liberties such as freedom of speech.

But the backlash against 'illegitimate' Trump voters is a sham defence of democracy that is just another attack on the independence of the *demos*. Anybody must have the legitimate right to protest against a president they don't support or to demand genuine democratic reforms. Demanding backroom deals among the political elite to overturn an election result you don't like, however, is potentially far more dangerous to the future of democracy than a President Trump.

However the arguments have been packaged, there is one underlying message of the backlash against Brexit and Trump: that 'too much' democracy is dangerous. The elites do not trust the mass of voters because they believe we are too unintelligent, misinformed and emotional to make the right decisions on important issues. And they do not really trust a lot of politicians either, who they think only win elections by pandering to the base appetites and instincts of the vulgar voters. An anti-democratic prejudice about lying 'populist' political demagogues and stupid voters is taking hold across the political and cultural elites on both sides of the Atlantic.

## Open debate about borders

The Brexit referendum vote was not a racist backlash but a revolt of the Others. It opened up the opportunity for a new kind of political debate about the future of our society, involving many who had previously been excluded from public life. Instead the reaction from the Clerisy and the political elites was to use it as an argument for even less democracy and openness in future: they want no more simple referendums on big issues, a bigger role for the courts in policing politics, official fact-checkers to sanitise 'post-truth' politics by restricting freedom of speech.

But more free speech and democracy, not less, is the best possible way forward, to give us a chance of addressing divided opinions, settling political differences and deciding which way to go. There is no point calling for unity and then demanding silence from dissenting viewpoints on any side. Democracy is about divided opinions and debate as to the way we shape our future.

An open democratic debate, for example, involving all opinions, represents our only chance of resolving a divisive issue such as immigration in UK and Western politics today.

In recent years, the prevailing view in British public life has been that anybody who raises the issue of immigration risks being branded a racist. This has suppressed debate on the question. Those deluded enough to imagine that was the same thing as winning the argument for open borders have had a rude awakening.

The first thing we need to do is clarify the issue through a clash of opinions rather than an exchange of insults. Concerns about immigration in the UK today generally have little in common with old-fashioned send-them-back racism. Instead mass immigration to the UK, especially from Eastern Europe, has become a symbol of the way that many people feel their world has been changed without anybody asking them. They have woken up to find that their communities are disintegrating, their traditional values trashed from on high. Some of their new neighbours may have their own native tongues, but the ones who really seem to speak a foreign language are the UK elites ignoring the UK's own 'ghastly people'.

In particular since the New Labour government of the late 1990s, mass immigration to the UK has been encouraged and organised from the top down, but without any public debate about its benefits or costs to society. Indeed any attempt at discussing immigration has been effectively barred as racist. Think of Labour prime minister Gordon Brown, unknowingly recorded dismissing a lifelong Labour voter as 'some bigoted woman' because she asked him about Romanian immigration on camera in the 2010 general election campaign.

Britain's borders have effectively been opened by the state, not as a consequence of governments or experts winning an argument for mass immigration, but instead by avoiding one and going ahead without public consent. In this context the immigration issue has become another symbol of the yawning gap between millions of people and the political establishment, of the absence of democracy and open public debate. You did not need to be a racist to revolt against that state of affairs.

Those who want a more liberal, open society would do better trying to win an argument for one than condemning those who disagree with them as xenophobes and thugs. The fact is that the precondition for any progressive policy on migration is establishing democratic control over borders – and then winning a democratic debate about the need to open them. The alternative of leaving it to the closed world of courts and Euro-commissions can only make matters worse.

Free speech and democratic debate are our best tools to tackle the political and cultural divide in our societies and arrive at some conclusions. Yet we live in a culture of conform-ism where the motto of the age is You Can't Say That, 'offensive' opinions are frowned upon or banned, and ideas that stray from the straight and increasingly narrow path approved by the Clerisy are ruled out-of-bounds.

One incident that highlighted this divide after the EU referendum came with the prosecution of the former England footballer Paul Gascoigne, ex-working-class hero, fallen national treasure and psychiatrically challenged alcoholic, for a race-hate crime. In 2015, during a desperate attempt to raise funds, the sad ghost of Gascoigne had staged An Evening With Gazza show in Wolverhampton. While on stage he said to his black security guard, 'Can you smile please, I can't see you.' For cracking this unfunny and insensitive joke, Gascoigne was hauled into court in September 2016 and convicted of the 'racially aggravated' offence of using 'threatening or abusive words'. Most telling were the threatening words used by the judge in sentencing Gazza. M'lud made clear that he was making an example of the faded star to send a message to others. 'We live in the twenty-first century,' the judge proclaimed. 'Grow up with it or keep your mouth closed!'[67] It is not necessary to defend what Gascoigne said in order to see that sort of censorious court order as no laughing matter.

You might interpret this sorry incident as evidence of the hidden epidemic of white working-class racism behind the Brexit vote. Or, alternatively, as a sign of the contempt with which the white working class and its old habits – such as telling naughty jokes – are held in high places. That helped to provoke the pro-Brexit backlash among millions who had had enough of being told to 'keep your mouth closed' by those who look down on them from the judges' bench of life.

### The big divide that matters most
The reactions against the referendum result and the Trump vote represent attacks on the idea of democracy and the equality of voters. This is now the big divide that really matters in Western politics, far more than that between Leave and Remain or Democrat and Republican. It is between the defenders and the opponents of greater democratic involvement in decisions that

affect us all. Whichever side you took in June in the UK or November in the US, this should be the issue to worry about and the cause to fight for now.

These issues create some odd alliances and strange bedfellows, cutting across the traditional political lines of Right and Left. In this as much else, we can draw inspiration from the great English democrat Tom Paine, a leading light in the eighteenth-century revolutions against tyranny in both America and France. When the great questions of democracy in his age arose, Paine insisted, the old party divides were rendered irrelevant: 'It is not whether this or that party shall be in or not, or Whig or Tory, high or low shall prevail; but whether man shall inherit his rights, and universal civilisation take place.'[68]

Without attempting to emulate Paine's rhetorical flourish, it seems reasonable to suggest that the stakes are similarly high today when deciding which side we are on in the battle over the future of democracy.

A couple of months after the EU referendum, *Wolf Hall* author Hilary Mantel recalled how some had tried to start a panic on 23 June about polling stations using pencils, raising fears that Leave supporters might have their votes erased afterwards. 'How we laughed!' she wrote. 'But then as soon as the result was in, millions signed a petition to rub it out and do it again. The bien-pensants suggested the result was not binding, but advisory – an opinion they would hardly have offered had the vote gone the other way.' Mantel compared the bitter Remain lobby to the 'army of erasers' she had encountered in Saudi Arabia, who dealt with things they didn't like – pork, Israel, women's equality – by simply removing mention of them from public life.[69]

Yet those millions not only exist, they have made their presence felt for almost the first time in memory. That Electoral

Reform Society report cited earlier also contained the insight, rather begrudgingly reported by the ERS despite its criticisms of the campaign, that its researchers had heard 'time and again' from people who felt the EU referendum was the first time their vote 'had truly counted'. People decided for themselves what the truth was about the EU, and made their own choice in defiance of whatever was flung at them by the political class.

It seems the EU referendum was a rare modern example of democracy that means something real, where votes 'truly count', because it is lived by the masses who take part in the moment. And that is why the democracy-fearing elites of the UK and the rest of the West could not handle the truth about what happened.

To mean something more, democracy has to be lived more than once every four or five years on election day, or a once-in-a-lifetime referendum, and we need to find ways to bring it alive. In the first place that means defending democratic principles against those who would tell us that some voters are more equal than others, and insisting that the problem is not 'too much' popular democracy but far too little.

When the alleged voice of liberal Britain, the *Guardian*, can produce an official post-referendum T-shirt that declares: 'Never underestimate the power of stupid people in large numbers',[70] encouraging its constituency to wear their prejudices with pride, perhaps we are not so far from a soft-pastel Farrow and Ball version of that painted barn wall in *Animal Farm*.

# 2

# Taking the *demos* out of democracy

'In the case of a word like *democracy*, not only is
there no agreed definition, but the attempt to make
one is resisted from all sides. It is almost universally
felt that when we call a country democratic we are
praising it: consequently the defenders of every kind
of régime claim that it is a democracy, and fear that
they might have to stop using the word if it were tied
down to any one meaning.'

– George Orwell,
*Politics and the English Language*, 1946[1]

Nobody with sufficient brain cells to write an 'X' in a box will
admit to being against democracy in the West today. Every
political leader signs up to the democratic process. There are no
mass rallies chanting, 'Give democracy the jackboot! Smash
universal suffrage!'

So what are we worrying about? Well …

While almost everybody in authority now pays lip service to
democracy, few seem keen to spell out what they mean by it.
The way that democracy is taken for granted, as a self-evidently

Good Thing in principle, can coincide with some highly selective ideas as to what it might mean in practice.

The rhetorical, ritualistic support for democracy belies the way that many reject its spirit. Politicians praising modern democracy has acquired the quality of old-fashioned Sunday school preachifying; something that makes the preacher and the choir feel worthy for the moment, but has little relevance to their real lives during the rest of the week. It might seem as if, like world peace and unlike puppies, democracy is for Christmas, but not for life.

For the rest of the time, in practice those in authority in democratic societies today seek to separate power from the people. The predominant sentiment is to limit the scope of democratic decision-making – by outsourcing important decisions to non-elected bodies, from the EU Commission to the courts, and by insulating the political class from popular pressure.

This rejection of the democratic spirit is creating a dangerous state of affairs. It widens the credibility gap between the ideals of democracy talked about by politicians, and the hollow democracy often experienced by the people in whose name it exists. Unless we can find ways to breathe new life into it, the 'Living Tree of Democracy' risks being reduced to dead wood, its shell still standing but dried out and lifeless inside.

## All democrats now?

We in the West are all democrats now, or so it seems. Fascism is definitely out of fashion, absolute monarchy is an absolute absurdity, and few would be willing to trade our low-growth economies for China's boom if it also meant we got to enjoy the benefits of their one-party Communist state. The only sworn anti-democrats are a few Islamist eccentrics who insist that voting for any mere human means winning yourself a seat in hell.

And democracy appears to be catching. By the year 2000, no fewer than 120 of the UN's 192 member-states considered themselves to be democracies. Of course, you can't always trust what people claim to see in the mirror. For example, those 120 self-styled democracies included the likes of the Chinese People's Democratic Republic, where an official responsible for the one-party local elections explained the limits of democratic freedom: 'There are 1.3 billion people in China. If they all expressed their opinions, who would we listen to?'[2] Oh, I don't know – how about 'the democratic majority'?

Nevertheless, the way that everybody wants a piece of the D-word demonstrates that even autocrats feel it necessary to dress up as democrats these days. Little wonder, given that most people on the planet will say they want to live under a democratic system. When the World Values Survey questioned 73,000 people in 57 countries which account for 85 per cent of the world's population, almost 92 per cent endorsed democracy as a good way to govern.[3]

And yet despite all of that apparent support, something is shifting beneath the surface of society that puts living democracy to question – not just in unstable developing states, but in its Western heartlands.

Democracy was in peril from the moment of its birth in ancient Greece. The word was first commonly used in Athens as an insult to those accused of encouraging mob rule, a dirty word to strike fear into the hearts of respectable property-owning citizens. In the subsequent 2,000 years, the only reason democracy was not under threat was that no semblance of it existed anywhere on Earth. Until a couple of centuries ago the idea of democracy was not merely forgotten but unimaginable.

Even in modern Britain, which has long prided itself as the home of 'the Mother of Parliaments', democracy was only introduced in piecemeal and begrudging fashion, more like a caricature wicked stepmother giving the children the odd spoonful of gruel to keep them from burning down the kitchen.

The British nation has recently been encouraged to commemorate the centenaries of the First World War, allegedly a prime historical example of us all being in it together. Yet as that war drew to an end in 1918, still only 60 per cent of British men and no women had the right to vote. Millions had been sent to fight and die for a state that denied them the most basic democratic freedom. The 1918 Reform Act finally extended the vote to almost all adult males aged twenty-one – but only to propertied women aged at least thirty. It would not be for another ten years, until 1928, that British women were given the same right to vote as men. In other words, universal suffrage in the UK is still younger than many older voters.

However, since the defeat of Nazism in the Second World War and the fall of Stalinism at the end of the Cold War, the political boot has appeared to be on the other foot. It became unimaginable for anybody who wanted to be taken seriously in the West to oppose democracy. They might as well have formed the Anti-Motherhood-and-Apple-Pie Party, with the demand to shoot puppies and sanctify paedophiles at the top of its manifesto.

Yet in these unusual times, something has begun to shift again. Anti-democratic sentiments long buried beneath the surface have been gradually breaking through the crust of our culture and gaining a greater public presence.

This is not a case of turning the clocks back or watching the fascist zombie rise from the grave. Western democracy today is coming under a strange new sort of challenge. In other times and places those trying to undermine the democratic process or thwart the will of the people might be kings and cardinals, right-wing dictators, army generals and their billionaire backers. And they would openly declare their hatred for mass democracy.

These days by contrast the leading critics of popular democracy are often not dictators and fascists but democratically elected politicians and liberals; not bigots of the Christian Inquisition or Islamist jihad but intellectuals, academics and experts from the higher reaches of Western culture. Denouncing the shortcomings of democracy has shifted from being the preserve of the reactionary right wing to being a fashionable cause on the liberal left.

What is more, these radical critics tend to launch their attacks, not as avowed anti-democrats like the oligarchs and aristocrats of the past, but in the name of democracy itself. For them, 'Defending democracy' comes to mean protecting the status quo – if necessary against the people. This is a form of mock democracy – demockracy – that turns the concept on its head.

It is high time we clarified what we mean by democracy, in order to take a stand for it against the new breed of 'democratic' elitists.

### The deplorable face of democracy

The first chapter showed the strength of the resurgent anti-democratic impulse first in the reaction to the Brexit vote in the UK's 2016 Euro referendum, and then in response to the election of Donald Trump as US president. But the new displays of antipathy towards practical expressions of democracy are not confined to a moment in British or American politics. Concern about the consequences of allowing 'too much' democracy has been crystallising in influential quarters across the West for some time.

It has become common in Western politics for parties to accuse their opponents of sinking into 'the politics of fear'. One political fear which appears to be growing more acceptable on all sides, however, is fear of the masses and what they might do if allowed to run riot through the ballot box.

We have seen it in the upper reaches of the European Union, with EC chief Jean-Claude Juncker warning against the threat of 'galloping populism' across the continent.[4] Galloping populism sounds like a nasty condition, until you realise that the 'virus' he is scared of is millions of people across Europe, from Hungary to the UK, voting for parties and policies of which the EU elites do not approve.

We can see it too in the growing concern in the US about too many people supporting the 'wrong' candidates and causes. The disdain felt towards the electorate was spelt out during the 2016 presidential election campaign, when Democratic (and they say irony is dead) Party contender Hillary Clinton told a Lesbian–Gay–Bisexual–Transgender fundraiser, 'You know, to just be grossly generalistic, you could put half of Trump's supporters into what I call the "basket of deplorables". Right? The racist, sexist, homophobic, xenophobic, Islamophobic – you name it. And unfortunately, there are people like that, and he has lifted them up.'[5] In her promiscuous use of the label 'phobic' to equate opinions she detests with some form of mental illness, Mrs Clinton seemed blind to her own deplorable display of what might be called *demos*-phobia.

Clinton's remarks caused some understandable outrage. Yet some leading Democratic commentators thought the problem was she had been too restrained in denouncing the evils of America's low-life voters, had actually not been 'grossly generalistic' enough in branding the phobics. The wrong word she used was apparently not 'deplorables' but 'half'. In the *Washington Post*, liberal academic Stacey Patton opined that, 'The only thing Clinton should have apologised for was her lowball estimate' of the number of racists and you-name-it-phobics.[6] *Slate* magazine's chief political correspondent Jamelle Bouie thought Clinton had been attacked 'for telling the truth' about American voters; he concluded that 'We're going to need a bigger basket' to hold all the millions of real deplorables.[7]

The notion of 'deplorable' voters being bad for democracy did not come off the top of Clinton's head. Nor is it really all about the awful Mr Trump. Clinton's outburst was the end result of a long-running high-level discussion about how too many American voters don't know what is good for them these days.

As far back as 1995, Christopher Lasch wrote in *The Revolt of the Elites* about how the 'simultaneously arrogant and insecure' professional elite now 'regards the masses with mingled scorn and apprehension'. For them, the term 'Middle America' had 'come to symbolize everything that stands in the way of progress', from 'family values' and 'mindless patriotism' to racism, sexism and homophobia. 'Middle Americans,' wrote Lasch, 'as they appear to the makers of educated opinion, are hopelessly shabby, unfashionable, and provincial, ill informed about changes in taste or intellectual trends … and stupefied by prolonged exposure to television'.[8] These 'ill-informed' Middle Americans are 'at once absurd and vaguely menacing' to the elites – not least when it comes to election time.

This obsession with the problem of US voters has been growing in liberal circles since the Republican George W. Bush was controversially elected president in 2000. Since they could not accept that this apparently inexplicable event really resulted from the political failings of the Democrats, it had to be down to the shortcomings of the electorate.

Historian and journalist Thomas Frank's influential 2004 book *What's the Matter with Kansas?* (published in the UK as *What's the Matter with America?*) captured the liberal exasperation with dumb working-class voters who desert their Democrat guardians to vote Republican. As Frank told a 2010 BBC radio documentary – revealingly entitled 'Turkeys Voting for Christmas' – he believed that the voters' 'preference for emotional engagement over reasonable argument' had 'allowed the Republican Party to blind them to their own real interests'. Frank went on: 'It's like a French Revolution in reverse in which the workers come pouring down the street screaming more power to the aristocracy.'[9] This focus on people allegedly voting against their own interests has gone further with worthy discussions about the 'low-information voter' (translation: low-intelligence voter) who does not choose by applying the power of reason but by using what leading liberal Arianna Huffington called their irrational, emotional 'lizard brains'.[10]

Long before Trump appeared on the electoral stage, these theories were all highlighting the inherent dangers of allowing political life to be influenced by the white trash, redneck hillbillies or other socially deplorable breeds of the 'American Idiot' that Green Day sang about. When he first stood for president in 2008, Barack Obama was recorded telling a private fundraiser that small-town Americans 'get bitter' about the world, which is why 'they cling to guns or religion or antipathy towards people who are not like them or anti-immigrant sentiment'.[11] Hillary Clinton, then Obama's rival for the Democratic Party nomination, condemned the remarks as 'elitist', and the future president apologised. Eight years later Clinton was telling a fundraiser of her own that those same Americans were not just bitter, but deplorables.

And once you see the mass of people in that way, the implications for democracy are evident: we don't want too much of it! Keep those basket cases away from the ballot box and any say in what happens to society.

Watching the rise of Trump in spring 2016, Andrew Sullivan, leading British-born conservative commentator on American politics, caught the mood of many in the US and won widespread plaudits for his argument: 'Democracies end when they are too democratic.'

For Sullivan, the danger is that the checks and balances on democracy introduced by America's Founding Fathers are no longer sufficient to 'cool and restrain temporary populist passions'. Aided by the internet, he protested, we are witnessing a collapse of 'reasoned deliberation' led by elites and its replacement by 'feeling, emotion' among the electoral herd. To prevent the horror of the system becoming 'too democratic', this leading voice of modern conservatism in the US called on all decent Americans not just to vote against Trump, the Republican candidate, but to rally to defend 'the elites' and 'the political Establishment' who 'provide the critical ingredient to save democracy from itself'. Sullivan concluded that: 'It seems shocking to argue that we need elites in this democratic age – especially with vast inequalities of wealth and elite failures all around us. But we need them precisely to protect this precious democracy from its own destabilizing excesses.'[12]

To save US democracy from itself, then, we need the elites to hold it in check. Which might sound a bit like that US army officer in Vietnam infamously explaining how 'It became necessary to destroy the town in order to save it'.[13]

The fashion for the elites to vote with their well-heeled feet against 'too much' democracy is most evident in the new emphasis on the importance of experts and technocrats, not just in offering advice to elected politicians but in running our advanced societies. This is an updated twenty-first-century version of an old argument against popular democracy. It seems that important issues to do with the future of the world are now far too complex and sophisticated to allow 'ordinary people' to have a say in decisions that will shape their lives.

After all, what do they know about the setting of interest rates or the pros and cons of intervening in the war in Syria? Better by far to leave such matters to a few top bankers or diplomats and generals. The fact that these managerial experts seem to manage to turn every financial and strategic crisis into a catastrophe at the expense of millions does nothing to alter their certainty that they know what's best for the rest of us.

As the deputy president of the European Commission, Frans Timmermans, informed an audience in Washington, US–EU trade relations are a matter for experts and political leaders to decide, and 'should not be left up to people who know everything about the way you slaughter chickens'. That would be the wrong sort of expertise, apparently.[14]

Alongside the elevation of the experts comes the emphasis on the perils of 'post-truth politics', which essentially seems to mean ignorant and gullible people trying to make important political decisions about Brexit or the White House without accepting what the expert elite tell them is true.

The consequences of this experts-know-best prejudice for democracy soon become clear. When new UK prime minister Theresa May gave a speech at the 2016 Conservative Party conference, criticising the consequences of central banks cutting interest rates and pumping billions of pounds into the economy, Bank of England governor Mark Carney was furious at her uninvited intrusion into his fiefdom. In response the governor spelt out that 'the policies are done by technocrats. We are not going to take instruction on our policies from the political side'.[15] The self-styled technocrats are not to be held accountable for what they do to the national economy by democratically elected politicians. (This is, of course, the fault of those same elected politicians who have freely outsourced such authority to the central bankers in order to avoid being held accountable. A state of affairs that, in the UK, is due to the actions of the New Labour government, which gave the Bank these powers to set interest rates while floating above demo-cratic politics.) When technocrats talk about 'independent' central banks, the key thing is that they should be independent of any democratic interference by the millions who pay the price for their policies. They seriously believe that they are the guv'nors of our economic lives.

We can see the same cool disdain for popular democracy flourishing in the 'Serious' section of the bookshop, where a leading European academic's latest work *Against Elections* is only trumped by an American professor's even more starkly entitled tome, *Against Democracy*.[16] Both do little more than rehash old prejudices. Yet both have been widely publicised and praised as capturing a current mood.

*Against Elections*, by Belgian academic David van Reybrouck, is subtitled 'In defence of democracy' – a fashion-able sleight of hand for a backdoor assault on that very thing. Van Reybrouck wants to replace some elections with the system of 'sortition' – filling political posts at random, by lottery. This, he says, must be more democratic, since it was practised by the ancient Greek inventors of democracy. Despite apparently trying to reinvent the chariot wheel, his proposals have been praised in high places, hailed as 'an idea whose time has come' in the book's cover blurb by South African Nobel Prize-winning writer J. M. Coetzee. Van Reybrouck, however, doesn't really want to leave his democracy-by-lottery to chance. He wants to choose leaders at random – but unlike the ancient Athenians, he does not truly believe that all citizens are equally qualified and anybody could do the job. He proposes instead that the citizens selected for public office by drawing lots would then be locked in a room with academics and other experts who can 'educate' the unthinking punters in how to think the correct thoughts and make the correct decisions. Thus we would get 'deliberative democracy' rather than the unthinking choices of the uninformed.

As often with these types, mention of a referendum in which everybody gets to vote on big issues really raises hackles. In a referendum, he complains 'you ask everyone to vote on a subject that only a few know anything about'; whereas in a deliberative system you allow a few people 'to consider a subject about which they are given all possible information'. The key question being: what information, given by whom?

Whereas 'a referendum very often reveals people's gut reactions', says Van Reybrouck, 'deliberations reveal enlightened public opinion'.[17] Enlightened, of course, by the elite educators and experts who have been spelling out which 'informed choices' the chosen few citizens should make. These re-educated, unelected members of a 'sortitive' body should then apparently be empowered to veto the 'wrong' laws passed by elected politicians who have been put there by the uninformed masses. Calling this plot to usurp the will of the electorate a 'defence of democracy' is a prime example of demockracy in action. The publishers rather gave the game away by illustrating the front cover of *Against Elections* with a picture of Donald Trump's infamous hair-do.

The widely discussed book *Against Democracy*, by Georgetown University professor Jason Brennan, is an even more brazen bid for shock-value publicity that nevertheless usefully spells out some of the assumptions which other thinking elitists might be too shy to confess so publicly.

Professor Brennan shamelessly endorses the prejudice that some voters are more equal than others. He divides 'democratic citizens' into three categories, given popular culture tags perhaps to show his students that academics can be cool. We are all apparently Hobbits – 'mostly apathetic and ignorant about politics' – or Hooligans – 'the rabid sports fans of politics' – or Vulcans – 'think scientifically and rationally about politics'.[18] No prizes for guessing which category the professor thinks should be beamed up into a position of greater-than-equal political influence.

Brennan calls for a system that would 'weigh some votes more than others, or might exclude citizens from voting unless they can pass a basic test of political competence'.[19] Of course, Brennan insists, as they all must, he is not really anti-democratic. He generously allows that we should probably keep the institutions of representative democracy, including elected assemblies. He simply proposes that these old-fashioned democratic practices be incorporated into a new system of 'Epistocracy', which would 'apportion political power, by law, according to knowledge or competence'. What sort of knowledge or competence, to be judged by whom? Simon Cowell and Judge Judy? Or perhaps a panel of Georgetown University professors?

Although the Professor claims that: 'The idea here is not that knowledgeable people deserve to rule – of course they don't', his proposals strangely all seem to point to that very conclusion. He suggests that an appointed 'band of experts' could veto 'inexpert' laws drafted by elected representatives. Or even that expert statisticians using the results of his political competence test could work out what the electorate '*would* support if only it were better informed'. His epistocracy could then enforce 'the public's enlightened preferences [which people did not know they preferred] rather than their actual, unenlightened preferences'. Otherwise known as giving people what they ought to want, whether they want or like it or not.

The argument of *Against Democracy* concludes that: 'Most voters know nothing, but some know a great deal, and some know *less* than nothing. The goal of liberal republican epistocracy is to protect against democracy's downsides, by reducing the power of the least-informed voters, or increasing the power of better-informed ones.'[20] Perhaps they should amend the Declaration of American Independence to read: 'All men are created equal, but only some can pass Professor Brennan's test of basic political competence', and while they are at it, alter the inscription on the Statue of Liberty to say 'Give me your tired, your poor, your huddled masses yearning to break free, your basically politically competent …'

We can see similar anti-democratic sentiments creeping into practical politics as well as academic thinking. Look at leading Western responses to events elsewhere, in the developing world. The self-proclaimed democrats who run our states will suddenly become fans of a 'pro-democracy' military coup if it overthrows an elected government they don't like, as happened in Egypt in 2013. That overthrow of democracy was backed by Western governments and given legitimacy by Western intellectuals such as *New York Times* columnist David Brooks, who opined that Islamists 'lack the mental equipment to govern' themselves. 'Incompetence,' declared Brooks, 'is built into the intellectual DNA of radical Islam', rendering radical Islamists 'incapable of running a modern government'.[21] The political problem of democracy is thus reduced to the failing 'mental equipment' and damaged 'intellectual DNA' of the Islamist government – and, by extension, of the voters who elected them. The logical solution then becomes not a political struggle to try to win the democratic debate, but sending in the men in white coats – or in the Egyptian case, the men in military uniforms – to restrain the mental patient.

More recently, when the Turkish military attempted to overthrow the elected Islamist government of President Erdogan in 2016, one US statesman spoke for many others by declaring his support for the coup plotters – in the name of democracy, naturally. Democratic Party Congressman Brad Sherman tweeted that 'Military takeover in Turkey will hopefully lead to real democracy – not Erdogan Authoritarianism'. On the US right, strategic analyst Lt Colonel Ralph Peters (retired) told Fox News that the military men behind the coup were 'the good guys' who America should support against Turkey's 'Islamist Fundamentalist authoritarian president', who he said wasn't really democratically elected (despite winning 52 per cent of the votes in Turkey's first direct presidential election). 'They're for democracy,' the military man said of the army commanders trying to overthrow the government.[22] Meanwhile Western governments held back to see if the half-cocked coup had any chance of success before coming out against it. Then they nervously watched the vengeful Turkish government impose its own crackdown on democratic freedoms, issuing empty rhetorical warnings to Erdogan while continuing to court him, because stability and control by the state is considered more important than people's liberties – especially in a Nato ally.

These and many others are all examples of a fashionable attitude that might be summed up as: 'I'm a democrat, but …' In essence it means: 'I'm all for democracy, but only in moderation. I think democracy is right, but don't want people voting for the wrong things. I know democracy is the best system in principle, but don't trust the worst people to decide issues in practice. I love democracy, but just can't stand the *demos*.'

This creeping democratic deceit, where everybody with influence is for it rhetorically, whilst increasingly acting to try to redefine and restrict democracy in reality, leaves us with the question: What are they so afraid of?

## The meaning of democracy

A clue to the answer lies in the origin of democracy. When the ancient Greeks created the portmanteau word *demokratia*, they surely can have had little idea how much trouble they were storing up for future generations of governing classes.

*Demokratia* had two constituent parts. Firstly, *demos* – 'the people'. This could either mean the population as a whole, as in Abraham Lincoln's declaration of 'Government of the people, by the people, for the people'; or *demos* could refer to the mass of poorer citizens, as opposed to the small wealthy elite, the oligarchs. (Either way, the definition of 'the people' was always restricted to adult male Athenian citizens; the mass of slaves owned by the citizenry were non-people, and the voteless womenfolk of Athens didn't count either.)

The second part of *demokratia* was *kratos* – meaning 'power' or 'control', derived from the Greek word for 'grip' or 'grasp'.

Taken together, then, *demokratia* meant the people not just having a say, but having political power; the masses not merely being consulted, but potentially grasping control over politics and society – including over the oligarchy. And it described a system in which that power would be exercised directly, by the citizenry themselves, not indirectly by a political elite of their representatives.

The Athens *demos* wielded its own *kratos* through the citizens' assembly, and also through the people's juries that made most laws; jurors were selected by lottery and left to legislate without the guiding hand of any autocratic judge.

Little wonder, then, that *demokratia* was heard most often as a boo-word meaning 'mob rule', the D-word invoked as a menacing bogeyman by the Athens elite and their supporters to condemn the advance of the upstart masses and frighten the oligarchical horses.

Little wonder either that, for the following 2,500 years, as the next chapter outlines, those in power in the West have had no interest in repeating the Athens experiment. First democracy was erased from history, denied and disregarded for two millennia. Since the modern age of mass society forced it back on to the political map, the ruling elites have sought to redefine democracy to mean something quite different and less dangerous than the original explosive brand of Athenian *demokratia*.

Since they were forced to acknowledge the virtues of democracy in principle, the practical concern of ruling elites has been to split the two constituent parts of *demokratia* – separating *demos*, the people, from *kratos*, power or control. They have sought to take the *demos* out of democracy and leave real power and control in the hands of the modern oligarchs, albeit often behind a democratic smokescreen.

## The great debate

A big early step in the sanitisation of democratic politics more than 200 years ago was establishing that the new Western democracy would not be direct democracy red in tooth and claw, like the Athenian model, but would be a paler version of representative democracy, whereby voters were granted a say in choosing which politicians would rule them.

In principle the idea of representative democracy sounds good. It all depends what we mean by the R- and D-words, how representative and democratic the system might really be. In most modern societies, from the first, the new system of representative democracy was intended to restrain democratic politics and popular power. Just as the elites wanted to divorce the *demos* (people) from *kratos* (power), now they sought to separate the system of representative government from too much democracy.

An important precedent was set after the American revolution of 1776 by the Founding Fathers of the US, who were clear that their newly independent nation would be a republic, but no Athens-style democracy. The system of elected government they developed was intended not only to represent the people, but more importantly to restrain and if necessary repress the democratic spirit. They introduced limited voting rights, electoral colleges, the Senate and the Supreme Court to, in the approving words of Andrew Sullivan, erect 'sturdy firewalls against democratic wildfires'.

A few years later, in response to the French revolution of 1789, a great debate broke out about the future form of government in Britain. On one side was Tom Paine, a key figure in both the American and French revolutions, who had urged the Founding Fathers to adopt a more radically democratic system; on the other, Edmund Burke, now seen as the father of Western conservatism. While Paine sought the overthrow of monarchy and the creation of a new society, Burke wanted to preserve the rule of tradition and erect those 'sturdy firewalls' to thwart 'democratic wildfires'.

Both men saw the need for a more democratic system of British government in response to the popular upheavals in France. And both favoured a system of representative rather than direct democracy as better suited to the size and complexity of their society. Paine acknowledged that it would be a poor reflection on the progress of humanity if they had to go back more than 2,000 years to ancient Athens to find a model for modern government.

There was a big difference, however, in the two competing ideas of what representative democracy might mean. Paine was clear that the elected representatives should be the agents of the people, effectively delegated by a rational electorate to carry out the decisions and choices which they had made. Burke, by contrast, was adamant that the representatives should be elected as an independent political class, to do as they thought best for the nation, not as they were told by the electorate.

There is nothing necessarily wrong with viewing elected MPs as representatives rather than delegated agents of the electorate. It is important that political leaders are able to lead and have minds of their own, proposing new ideas for which they will then be accountable to their electorate. It is also useful for elected politicians to be able to draw on the collective wisdom and experience of parliament. For the conservative Burke, however, it was clear that the point of this system of representative democracy was to separate the 'representative' from the 'democracy', insulating parliament from popular demands and allowing MPs to act as a restraining influence on the passions of the people.

In his *Reflections on the Revolution in France*, Burke wrote that: 'When the leaders choose to make themselves bidders at an auction of popularity, their talents, in the construction of the state, will be of no service. They will become flatterers instead of legislators; the instruments, not the guides, of the people.' As Burke the MP had told his few enfranchised Bristol voters in 1774, 'Your representative owes you, not his industry only, but his judgement; and he betrays instead of serving you if he sacrifices it to your opinion.'[23] Representatives were to be the people's 'guides' and 'judges', elected to serve the interests of the Crown and the state rather than the opinion of the small electorate – never mind that of the great disenfranchised populace.

In that late eighteenth-century era, Paine's revolutionary allies succeeded in overthrowing the old orders in both America and France. However, it was Burke's conservative view that won out in the longer-term battle over the nature of Western democracy.

Representative democracy came to mean the rule of a political class, periodically elected by those more or less begrudgingly granted the vote, but not accountable to their electorate except in a formal sense on election day every four or five years. (At that time in any case, few MPs were even elected by popular votes, and those that were had only to contend with a small electorate of property holders.) Paine's vision of an enlightened democracy where representatives would act in unison with and on behalf of a politically engaged people has rarely existed anywhere in the West.

## The fear of referendums
The redefinition of representative democracy to mean a system of rule by a political elite became accepted as the norm. Any dalliance with more direct forms of popular engagement could

then be denounced as a threat to democracy. This helps to explain the long-standing antipathy in UK and European political circles to the idea of direct democracy via referendums, which are seen as a challenge to the authority of parliamentary democracy. Indeed they can be – and what, we might ask, is automatically wrong with that, if parliamentary sovereignty clashes with popular sovereignty?

All shades of opinion in the British political class are generally hostile to the very idea of referendums. When the Tory opposition called for the New Labour government to hold a referendum on the EU's Lisbon Treaty in 2005 (see chapter 4), foreign secretary David Miliband summoned the authority of former Conservative prime minister Margaret Thatcher, reminding his Tory opponents that she had once denounced the referendum as 'a device of dictators and demagogues'. At the same time Andrew Duff, leader of Britain's Liberal Democrats in the European Parliament, made clear that: 'The plebiscite is a form of democracy, possibly suited for revolutionary circumstances, but completely unsuited for informed and deliberative decisions on complex treaty revision.'[24] Whether they consider referendums to be the tools of dictators or revolutionaries, our political class is normally united in the belief that they are not suitable for making big decisions in a democracy.

No doubt a system based on government by continual referendums would be unwieldy and impracticable, but where the political elite has become so unrepresentative of the majority – as for example on the UK's attitude to the EU – then a referendum can present a usefully direct device for expressing the democratic will and breathing new life into politics. The 2014 referendum on Scottish independence had a similarly rejuvenating, if short-lived, effect on public debate. It is precisely the big and decisive character of the issues at stake that makes them a suitable case for a direct appeal to the electorate in the form of a referendum. And whether you approve of referendums or not, once one has been held, democracy surely demands that the popular decision be implemented.

The dislike of the British political class for referendums is not based on a love of parliamentary democracy, but on fear and loathing of the *demos*. It is interesting that those who loathe the referendum will often use, in a derogatory sense, the alternative word 'plebiscite', from the Latin *'plebs'* – the common people.

And it is not only in Britain that European reactions to recent referendum results reveal this powerful disdain for expressions of direct democracy. In July 2015, Greece held a referendum on whether to submit to the latest austerity measures imposed by the Troika of the European Commission, European Central Bank and IMF. The radical left-wing Syriza government called on the people to reject the package, prime minister Alex Tsipras telling an Athens rally: 'I call on you to say a big "no" to ultimatums, "no" to blackmail. Turn your back on those who would terrorise you.'[25] The Greek public followed his lead – a resounding 61 per cent voted *'Oxi'* ('No!'). Barely a week later, the same Tsipras bowed to the 'terrorist' Troika's blackmail demands and signed the very measures the referendum rejected, from VAT rises to public-sector pay cuts. The opinions of a few Euro-technocrats had outweighed those of millions of Greek voters, much to the approval of the EU elites.

Elsewhere in the EU in 2015, Ireland was cheered to the rafters after 62 per cent voted 'Yes' in its referendum on legalising same-sex marriage, which apparently proved what an enlightened European nation Ireland is. No relation, presumably, to the Ireland castigated as a backward outpost after it rejected the EU elite's Lisbon Treaty in a 2009 referendum. On that occasion, the EU made the Irish hold another referendum to ensure they got the 'right' result. By contrast there was no question of rerunning the same-sex marriage referendum which produced the result that the elites agreed was correct. Ireland had apparently passed its entrance exam.

## From apathy to populism

The Western system of representative democracy has more recently evolved into an increasingly unrepresentative one. This has involved two important trends. First, the concentration of greater power and authority in the hands of unelected and unaccountable bodies, whether that means the courts, commissions of experts, public inquiries or the institutions of the European Union. And second, the professionalisation of politics and increased estrangement of the political elite from the people who it is supposed to represent.

An unpopular European political class that has lost (insofar as it ever had it) public trust and lacks the public authority to rule in its own name has become keen to 'outsource' important decisions to unelected, unaccountable institutions. This is intended to give governments some protection from popular discontents, allowing them to pass the buck to the judiciary or the EU. Whatever good it might do the government, however, it can do nothing but harm to democracy. It might suggest that our system of government is already far closer to an elitist 'epistocracy' than some would like to admit.

As chapter 4 argues, the European Union has often played this role of bypassing democracy for member-states. It has been less a case of Brussels imposing its will on reluctant national governments, and more of those governments willingly handing the EU the authority to implement decisions without the awkward need to seek the democratic consent of their people.

Nearer home, the UK judiciary has become an increasingly powerful interventionist force in the political life of our society. In setting up the new Supreme Court and incorporating the sweeping legal powers of the European Convention on Human Rights into UK law (via the 1998 Human Rights Act), the New Labour governments of Tony Blair and Gordon Brown promoted the judges into a position of even more magisterial influence in public life.

The rise of the judicial elite as arbiter not only of individual legal cases but of wider issues facing the whole of society undermines the very idea of representative democracy. No British judge is elected to his or her position on the bench, nor can they ever be removed from it by the electorate. Judges 'represent' only the power of the Crown, as symbolised by the royal slogan, 'Dieu et Mon Droit' – 'God and My Right' – which hangs over their court. They are accountable for their decisions to nobody other than their own conscience and that of more senior judges.

For some, the more interventionist role of the courts might provide a welcome counterbalance to what they claim is the 'electoral dictatorship' of the legislative and executive arms of government. For others, however, it poses the danger of what critics call 'judicial supremacy' or even 'judicial dictatorship'.[26] And unlike the 'dictators' of Whitehall and Westminster, those sitting in the Supreme Court cannot be removed via the ballot box.

The anti-democratic implications of 'judicial supremacy' were well illustrated after the UK's referendum on EU membership. Top lawyers acting on behalf of a secretive group of financiers calling itself 'The People's Challenge' (*sic*) went to the High Court in London to ask senior judges to rule that, in spite of the referendum result, the government could not trigger Brexit without the approval of both houses of parliament. Thus three high court judges were put in a position effectively to overrule the wishes of 17.4 million voters, 'if it please your lordships'. The high court's judgment in favour of the anti-Brexit applicants prompted a furious reaction from sections of the press. Yet it should surely have come as no surprise, in a society where the judiciary has been empowered to act increasingly as a law unto itself on public issues that should be none of its business.

An often-overlooked corollary of the rise of the judges is the UK authorities' obsession with setting up judge-led public inquiries into anything and everything. In recent years there has been an explosion of public inquiries into issues as diverse as phone-hacking and press regulation, the 2003 war in Iraq, the Jimmy Savile scandal, alleged Establishment child abuse cover-ups, and many more. (Sir John Chilcott, who led the inquiry into the Iraq war, is a former civil service mandarin and an appointed member of Her Majesty's Most Honourable Privy Council – the type of top state official who's no more publicly accountable than a senior judge.)

One thing they all have in common is that they begin from a crisis of public mistrust of powerful institutions – be that parliament, the press, the police or the BBC – and try to bypass the problem by handing it over to judge-led inquiries that are still supposed to inspire some public confidence. As one academic study of these trends has it, 'Major inquiries draw upon judicial independence to restore political authority.'[27]

The question this poses is: independent of what? What these judicial inquiries are truly independent of is public accountability and democratic debate. They take issues that ought to be burning questions of public controversy – such as the causes of wars, the freedom of the press, or society's attitude to children – and remove them from the democratic arena. Instead they are locked away within smoke-free inquiry rooms for months or even years on end, while the experts hold forth until the presiding judge or chair eventually issues a verdict. There seems no more sure way to drain the life out of democracy than that.

Setting up a public inquiry, or simply demanding one, has become a substitute for democratic debate and actually doing something about a problem for government and opposition alike. During the Conservative–Liberal Democrat Coalition government, one exasperated writer in the Labour-supporting *New Statesman* asked: 'How many independent inquiries has Labour called for?', noting that the Opposition party's 'default response to scandal is, increasingly, to demand an independent inquiry'.[28] For the Opposition, demanding an inquiry is a way of putting pressure on the government, without having to argue for it to do anything specific or real. For the government, responding to such demands allows it to look concerned, but without taking political action. On both sides, the obsession with public inquiries led by judges who are independent of the public detracts from proper democratic debate and accountability.

The second trend that has rendered democracy less representative still has been the increased estrangement of the political class from the people.

In recent decades the distance between the mainstream political parties in most Western societies grew narrower, giving electors less genuine choice between the different wings of a managerial ruling class. Meanwhile the gap between the entire political elite and the people grew ever wider, as the politicians and their media allies retreated into the closed oligarchical world of the Westminster bubble or Washington Beltway.

The Cold War ended with the collapse of the Soviet Union a quarter of a century ago, and the political struggle between Left and Right that had defined public debate over the previous two centuries effectively ended with it. We were faced instead in most Western elections with a bland system of managerial politics and a choice between interchangeable candidates that had little more meaning than the choice between near-identical banks or supermarkets on the high street.

Western political elites have all looked increasingly like members of a modern oligarchy, the differences between the parties far less striking than the similarities between the politicians. In Germany, for example, the two great competing parties of post-war politics, the Christian Democrats and Social Democrats, have even joined together in a coalition government.

This trend was accelerated by what might be called the professionalisation of public life – the process whereby top figures in politics, government and the media emerge from a similar educational and career path, forming a professional layer quite apart from the rest of society.

Thus it became the norm for UK MPs to 'train' for a job in parliament by being political researchers or advisers to other MPs, trade union or local government officials, or via a short career in the top echelons of the media or public relations, or some international charity or global institution. They would then often be 'parachuted' into a parliamentary seat, to represent local people with whom they might well have no connection and little or nothing in common. Typically, former Labour leader Ed Miliband, son of a radical north London academic, somehow became MP for the old industrial town of Doncaster, in Yorkshire – the first Labour MP to occupy that seat without first having been a coal miner.

The death of the politics of choice and the professionalisation of public life accelerated the estrangement of the political class from the populace. The result has been to accentuate the public's drift away from politics. In recent decades the pattern across the West has been one of falling voter turnouts and memberships of political parties and less loyalty to the established parties.

For many in the ruling elite, this is not necessarily a problem in itself. There has long been an acceptance that what is called voter apathy, but usually means an alienation from politics, can be a good thing if it keeps the wrong sorts of people out of the voting booth. As the US political scientist Samuel Huntington expressed it back in the unstable 1970s, an 'excess of democracy' can be a problem for those in power. Instead, thought Huntington, 'the effective operation of a democratic political system usually requires some measure of apathy and non-involvement on the part of some individuals and groups'. The more antagonistic these individuals and groups were to the political status quo, the more their 'non-involvement' in the allegedly 'democratic political system' would be welcomed.[29]

However, while a bit of grumbling public apathy is one thing, growing antagonistic support for dissident parties is something else altogether. In the past few years the reaction against what has happened to politics has encouraged a drift towards apparently anti-establishment parties and candidates on the Right and the Left, from Greece and Italy to Poland and Hungary, and even from the UK to the US. This has sent a tremor of trepidation through the old political elites and technocratic authorities.

This explains the ferocious backlash against what they call 'populism' across European and US politics. As with 'democracy', a glance at the origins of the word gives a clue to what the real concerns might be. Like democracy, populism comes from a word meaning the people – in this case from the Latin, *populus*. And like the word democracy in the past, 'populism' is most often used as a term of abuse these days. Ask what the critics mean by it, and the Cambridge English Dictionary will tell you, for example, that it is a 'negative' description of 'political ideas and activities that are intended to get the support of ordinary people by giving them what they want'. Giving ordinary people what they want? We can't have that, can we?

Back in its nineteenth-century youth, 'populist' was a badge worn with pride by anti-establishment political movements. Now it is a term of abuse, a label attached to parties and policies by the establishment itself. As political scientist Ivan Krastev asks, 'Who decides which policies are "populist" and which are "sound"?'.[30] That decision is apparently the preserve of the anti-populist elites of European and American politics.

The political mainstream is keen to try to discredit the revolt against its values by branding any non-conformist ideas as 'populism'. It appears as if the elites are content to be thought of as 'unpopular', if the alternative is to court popularity among the untouchable 'deplorables'.

Faced with the 'populist' backlash and wider disaffection, the West's defensive elites have also sought to offer some evidence that democracy is in a healthy state and that the people are still engaged with it. We have been presented with desperate attempts to talk away the crisis of democracy and show that our increasingly unrepresentative system is really responsive and in touch with the concerns of the population. The buzzword is 'consultation' rather than representation. These only end up further degrading the meaning of genuine democratic involvement.

One example is the obsession with online political engagement. The blossoming of social media websites such as Twitter has been hailed as the re-energisation of democratic debate. The profusion of online petitions is praised as a new form of public participation; in the UK, the political class has sought to forge a new virtual link with voters by vowing that any petition which gains more than 100,000 online signatures will be discussed in parliament. The impression they want to give is that democratic engagement is alive and well online. These efforts to sell the virtues of online democracy have been greatly aided by a layer of lobbying activists on consumer or environmental issues, who like to present the media-focused protests of a few professional zealots backed by a push-button petition as a new brand of 'mass democracy in action'.

The online world may indeed present new opportunities for political action and engagement. But this attempt to elevate the meaning of casual internet activity only further lowers the value of voting and democratic participation. The push-button democracy of endorsing an online petition or giving a few pounds in response to advertising by a party or campaign is hardly a very meaningful political act. To claim that it is the moral equivalent of real democratic participation only makes the latter look more dead than alive. It endorses a type of ersatz democracy which, like the ersatz coffee made of acorns that soldiers drank in the trenches, might be presented as looking like the real thing, but is an unsatisfying substitute unlikely to make you keen to do it again.

## From right to left

The extent of the new dangers becomes clearer when we ask: who is genuinely on the side of greater democracy today?

In the twenty-first century, anti-democratic arguments are no longer the preserve of old-fashioned right-wing reactionaries and aristocrats. On the contrary, their most eloquent and earnest champions are now to be found on the liberal or even radical left of the intellectual and political spectrum. This is one of the worst things about the new anti-democratic turn for those of us who have long seen the demand for more democracy as a left-wing cause.

From the origins of the political divide between Right and Left in the French revolution of 1789, the struggle for greater democracy was always associated with the latter. At first, in the National Assembly, supporters of monarchy and the king stood to the right of the president, while supporters of the revolution stood to his left. When the National Assembly was replaced by the revolutionary Legislative Assembly in 1791, the more radical 'innovators' stood to the left while the more conservative 'conscientious defenders of the constitution' were on the right. The demand for greater democratic change and popular control largely belonged to the Left through many tumultuous struggles that followed, at least until the tyranny of Stalinism.

Yet today, the Western Left has largely abandoned its championing of democracy in favour of more state control and technocratic politics. This reflects the extent to which, after the political defeats of the twentieth century, the Left has lost touch with the mass working-class constituency it once aimed to represent, and instead invested its trust in an enlightened liberal elite.

One feature of the contemporary Western Left is its obsession with the influence of the mass media. As discussed in a later chapter, criticising the mass media and the popular press is largely a coded way of attacking the mass of the populace. The ostensible target is Big Media, but the real one is the Big Public, the people deemed sufficiently gullible and ignorant to be easily led astray by wicked press barons and TV or web moguls.

This argument is personified by somebody like the American professor Noam Chomsky, a hero of the international Left who has been named as the world's top public intellectual. Chomsky focuses on how the power of the mass media 'dulls people's brains'. He refers to the masses as that '80 per cent of the population whose main function is to follow orders and not think', typified by the beer-drinking working stiff 'Joe Six Pack' whose exposure to the corporate-controlled mass media has 'reduced [his] capacity to think' rationally, resulting in totalitarian-style 'brainwashing'.

The titles of Chomsky's best-selling books – *Necessary Illusions: Thought Control in Democratic Societies*, and *Manufacturing Consent: The Political Economy of the Mass Media* – give his game away. It's as if our very thoughts and opinions are controlled and manufactured by the big media.

It might all sound very radical and anti-corporate, yet the underlying message echoes that of conservative anti-mass writers of a century ago. Chomsky's book title, *Manufacturing Consent*, comes from a line by Walter Lippmann, whose 1922 book *Public Opinion* is a classic of the old anti-democratic elite.

Lippmann bluntly argued that much of the American electorate was 'absolutely illiterate', 'mentally children or barbarians' who made easy prey for 'manipulators' in politics and the media. Lippmann believed that it should be left to the better-educated classes to decide what was best for the rest.[31]

Today's radical left thinkers might not use Lippmann's overtly elitist language (though Chomsky has repeated Lippmann's view of the American public as a 'bewildered herd'). But the message is much the same: that the malleable masses are easy prey for media manipulators who can brainwash them to 'follow orders and not think'. As Brendan O'Neill wrote in *The Australian*, after the good professor won the

Sydney Peace Prize, 'The reason today's PC chattering classes adore Chomsky is because he has made it respectable again to pity and/or fear the mob.'[32]

As with many who are suspicious of popular control, the sight of an exercise in direct democracy through a referendum sends serious shivers down the spine of our new left elitists. The Slovenian philosopher Slavoj Zizek has become a cult hero among many young leftists in recent years, seen almost as the last European intellectual standing for fundamental political change.

In response to the Brexit referendum result, however, the radical Guy Fawkes mask slipped to reveal a different face of Zizek beneath. 'You know,' he told openDemocracy days after the referendum, 'popular opinion is not always right. Sometimes I think one has to violate the will of the majority.' He declared that 'Direct democracy is the last Leftist myth'; rather than an impractical and simplistic referendum, the great radical thinker would prefer 'the appearance of a free decision, discreetly guided' by a 'discerning elite.'[33] It seems today's left would rather put their trust in the discretion of an elite than in the unruly and unpredictable 'will of the majority'.

## Majority wrongs and minority rights

The new leading role of left-wing thinkers has helped to change the language of the attack on democracy. The 'tyranny of the majority' is today more likely to be depicted as a threat to the minority rights of identity groups, rather than to the property rights of the rich and powerful. But the 'progressive' change in tone does not alter the reactionary consequences of rejecting greater democracy in favour of elite control.

In *Athens on Trial*, the historian Jennifer Tolbert Roberts notes how the recent elevation of leftish identity politics has not only affected how democracy is seen in the present, but has also changed the historical critique. In the past, establishment historians attacked ancient Athens for allowing the mass of male citizens to usurp the power of the elite oligarchs; these days radical historians denounce the citizens of Athens for their ill-treatment of women and slaves. 'The poor Athenians, it seems, cannot win,' says Tolbert Roberts. 'Once censured as crass levellers, they now find themselves under fire as closet aristocrats.'[34]

Democracy has indeed posed an historical threat to the powerful and privileged (albeit one which, as we shall see, they have often managed to negotiate). There have also been circumstances in which majority rule has conflicted with minority rights. Most importantly, however, more democracy remains the best hope for achieving real freedom and equality in society.

Liberty is the precondition for fighting for equality, and democratic rights are our most fundamental freedoms. Minority rights have advanced furthest as part of the struggle to wrest universal democracy from the ruling elites, not in a culture war against the majority. Reclaiming democracy as a radical cause for all those who want to see political and social change is one of the big challenges we face.

One old notion that has come back into intellectual fashion is the idea of a conflict between mass democracy and individual liberties. It is an unfortunate fact of history that some of the most forthright intellectual advocates of the individual right to freedom of speech, from Socrates to John Stuart Mill, have been far less keen on the democratic rights of the masses.

In his recent book *Against Democracy*, US professor Jason Brennan seeks to take on the mantle of Mill and co. by counter-posing democratic rights to civil liberties. For Brennan and his ilk, 'The right to vote is *not* like other civil liberties, such as freedom of speech, religion or association.' This is essentially because, in contrast to individual liberties, democracy allows the majority to 'impose their decisions on others'. In Brennan's contemptuous view of the majority, if 'most voters act foolishly, they don't just hurt themselves. They hurt better-informed and more rational voters, minority voters', and many others. By contrast, claims the professor, an 'epistocracy', in which those 'better-informed and more rational voters' were given more power, and enlightened experts could veto the actions of elected representatives, 'would protect and promote liberal freedoms better than democracy does'.[35]

Indeed such elitist notions are already put into practice via the courts. The nine justices who sit on the US Supreme Court are widely seen as the protectors of constitutional rights such as freedom of speech against intolerant law-makers and public opinion. In the UK, high court and supreme court judges are increasingly assuming the role of interpreters and protectors of individual rights via the Human Rights Act, which incorporates the European Convention on Human Rights (ECHR) into British law.

This raises a fundamentally false counter-position between liberal freedoms and democracy. Individual liberties, from the emancipation of slaves to women's rights, have been won and defended as part of the broader historical struggle for a more free and democratic society. How else does anybody imagine an individual would have had the strength to wrest their rights from the powers-that-be, and hang onto their liberties, other than as part of a wider collective fighting for more democracy?

The likes of Professor Brennan may be content to entrust his civil liberties to the goodwill of a few 'better-informed' individuals, empowered to overrule democracy. But what would be the meaning of 'freedom of speech, religion or association' in such an undemocratic system? What could it mean to claim your free speech was protected, if you were denied an equal say in shaping the future of society? You could talk all you liked but nobody would take any notice.

That sounds like the sort of 'free speech' that might appeal to a hermit meditating in a cave, or perhaps a professor who fancies a seat on an elite epistocratic committee. But it would be of no use to anybody fighting for a freer society.

The real empowerment of unaccountable judges, as supposed protectors of our civil liberties, is a sharply two-edged sword. It also gives them the power to restrict or override those rights if they so choose. In recent decades, for example, the US Supreme Court has generally been seen as a champion of free speech upholding the First Amendment to the Constitution, reflecting the more liberal times. Yet in past periods of social tension, such as the anti-communist scares that followed both world wars in America, the Supreme Court refused to uphold the First Amendment rights of those individuals it viewed as subversive or 'un-American'.

America's version of democracy, as Larry Diamond from the Hoover Institute at Stanton concedes, involves 'having 9 unelected justices with lifetime tenure and no political account-ability to anyone but themselves decide such basic questions as when a woman can have an abortion or where a child can go to school'.[36] The fact that most Americans might have supported the Supreme Court's past liberal judgments on abortion or segregated schooling does not alter the fundamentally undem-ocratic nature of that system, or the potential dangers of entrusting the liberties of more than three hundred million American citizens to the verdict of nine justices appointed by the President and approved by the US Senate.

These days the top courts of the EU and the UK are also often depicted as the best defenders of individual liberties, by enforcing the ECHR and Human Rights Act. However, that power too can cut both ways. The Human Rights Convention, for instance, enshrines the right to freedom of expression. Yet the ECHR also spells out the grounds on which that right can be legally restricted and penalised 'in a democratic society'. They include such broad terms as 'in the interests of national security … or public safety' and for 'the protection of health or morals'.

As I noted in *Trigger Warning*, my book in defence of freedom of speech, these 'sound like the sort of catch-all excuses for restricting free speech beloved of dictators down the decades. It is the restriction of speech in the name of freedom. And it is ultimately up to the learned judges of the UK and European courts, of course, to decide just how much liberty to allow'.[37] Those UK judges have recently proved willing to use the Human Rights Act further to restrict the right to freedom of speech and of the press and the public's right to know, on the pretext of upholding the right to privacy. We entrust them with our democratic liberties at our peril.

Of course it is possible for a democratic majority to endorse a decision that goes against minority interests. But Brennan's belief that democracy basically means 'most voters act foolishly' owes more to snobbish prejudice than political science. The alternative perspective is that democracy is the best hope for all of our freedoms. We can become more liberated, autonomous individuals by becoming part of something bigger than ourselves, transcending the narrowly personal and getting more involved in public life as part of living democracy.

## Living democracy

What is to be done about this hollowed-out demockracy we are offered today? On one hand it appears that democracy is taken for granted, as if it has always been here and always will be. On the other the real meaning of democracy in Western society today remains unresolved.

Democracy has always been more than a word or a nice idea. A dynamic version of democracy can harness people's creative energies. It will encourage people to participate in public life and take responsibility for their decisions, because what they say and do really counts. What matters is not just how democracy is talked about. It is how it is lived by the citizens of a democratic society.

It is a fundamental element of our humanity to be sovereign citizens with a say over our own lives and societies. Opening up the world to democratic participation gives us the opportunity to influence events outside of ourselves and to live with the consequences of our own actions, rather than simply the actions of our rulers.

Democracy is discussed today as the best or least bad system of government. In the classical view, however, it was about the way a society was founded, rather than just a political system. Democracy was about a view of humanity, seeing citizens as fit and able to participate on an equal basis. It was not just about the ability to place a cross in a box, but about the fundamental values for which a democratic society, and its members, should stand and fight.

Ancient Athens used a system of sortition to select its public officials and jurors by ballot, on the assumption that each of them would be able to do the job as well as others. In the words of the twentieth-century black writer and activist C. L. R. James, they believed that 'Every cook can govern' (just so long as that ancient Athenian cook was a male citizen).

One of the few surviving pro-democracy texts from ancient Greece, the funeral oration by the democratic leader Pericles for the dead of a great war, captures the spirit of Athenian democracy to do with public participation:

'No one, so long as he has it in him to be of service to the state, is kept in political obscurity because of poverty. And just as our political life is free and open, so is our day-to-day life in our relations with each other ...: we do not say that a man who takes no interest in politics is a man who minds his own business; we say that he has no business here at all.'[38]

When democracy was reawakened in the West 2,000 years later, that spirit once more tried to take hold across society. Visiting the New World in the 1830s, to write what would be his classic work *Democracy in America*, the French aristocrat Alexis de Tocqueville was struck by the unprecedented energy of the republic and its people: 'Democracy does not give the most skilful government to the people, but it does what the most skilful government is powerless to create; it spreads a restive activity through the whole social body, a superabundant force, an energy that never exists without it ... [It] can bring forth marvels. These are its true advantages.'[39]

Tocqueville was clearly describing the American system, not as a dry governmental process, but as a living, pulsing democratic society.

Dynamic democracy has proved to be an engine of human progress and emancipation. The spread of democratic freedoms in the modern era has often accompanied the rise of economic dynamism and wealth. But it is about much more than economics – indeed there are other recent examples of economic growth alongside the stagnation of freedoms.

A flourishing of democratic participation brings a greater sense of our humanity that has also led to the flowering of arts and culture. Ancient Athens is famed not only for its democratic politics, but also for its historic advances in everything from sculpture and architecture to drama, notably in the 'golden age' of Pericles. Centuries later the new republics of Italy became the cockpit of an explosion of the arts in the Renaissance. In the subsequent age of the Enlightenment, the same growing sense of individual autonomy that sparked the demand for freedom also gave greater rein to the arts and literature. The way that the struggle for liberty and democracy can ignite human creativity has often been evident since.

Five hundred years ago, the Italian Renaissance philosopher Niccolò Machiavelli was already clearer than many appear today about the relationship between popular government and progress: 'Further, we find that those cities wherein the government is in the hands of the people, in a very short space of time, make marvellous progress, far exceeding that made by cities which have been always ruled by princes; as Rome grew after the expulsion of her kings, and Athens after she freed herself from Pisistratus; and this we can ascribe to no other cause than that the rule of a people is better than the rule of a prince'.[40]

That reflection comes from 'The Multitude is Wiser and More Constant Than a Prince', which Machiavelli wrote as a chapter in his Writings on Livy. 'As for making judgments,' he declared, 'when the people hear two opposing speakers of equal skill taking different sides, it is only on the rarest occasion that it does not select the best opinion and that it is not capable of understanding the truth it hears.'

The people could certainly go astray, thought Machiavelli, but were always open to reason, unlike those who thought themselves born to rule: 'For a turbulent and unruly people may be spoken to by a good man, and readily brought back to good ways; but none can speak to a wicked prince, nor any remedy be found against him but by the sword.' And the fact that the populace could so often be criticised and traduced in public was itself a sign of the superiority of popular government over one where nobody dares to criticise the prince: 'The prejudice which is entertained against the people arises from this, that any man may speak ill of them openly and fearlessly, even when the government is in their hands; whereas princes are always spoken of with a thousand reserves and a constant eye to consequences.'

Today, Machiavelli is most often recalled as the author of *The Prince*, a study in how to manipulate the masses in 'Machiavellian' fashion that many compare to modern *House of Cards*-style politics. It is striking to be reminded how public-spirited and open-minded this five-centuries-old humanist appears, as a supporter of government by the 'turbulent and unruly people', who he deemed quite capable of judging all opinions and 'understanding the truth it hears'. It presents a remarkable contrast to the elitist misanthropy so prevalent in our modern supposedly democratic culture, dominated by 'the prejudice which is entertained against the people'.

## No guarantees

Real democracy is rarely the easy option and there are no guarantees of good government. Once the democratic genie is out of the bottle, he may not obey your orders or grant your wishes. Any historical critic can find examples of electorates making dangerous and damaging choices, from the civilised Athenians taking a democratic vote to execute their greatest philosopher, Socrates, for talking out of turn, to millions of Germans voting for Hitler's Nazis in the cultured heart of twentieth-century Europe.

No doubt it is also true that an exercise in more direct democracy, such as a referendum that boils down major issues to a simple question, can provide an opening for unscrupulous politicians to seek to manipulate public opinion. As Zizek says, 'you know, popular opinion is not always right'. But the possibility that exercises in either direct or representative democracy can sometimes go wrong is no argument against it.

Of course genuine democratic debate is inherently risky. The 'safe' alternative would be a fixed election and a debate in which one side was banned from speaking. The hard truth about democracy is that it involves indivisible liberties for all, and everybody gets the same vote.

Anybody inclined to try to limit the democratic rights of others who disagree with them is playing a potentially dangerous game. Once you treat democracy or liberty as a selective privilege to be handed down, rather than a universal right to be demanded, your own rights are also at risk. Those who would prefer to live by the court order, the Royal Prerogative or the Euro commissioners' ruling can ultimately perish by those undemocratic instruments, too.

You do not get to pick and choose which bits of democratic decision-making you want to keep, or to send back those outcomes that are not to your liking. The danger of democracy, however, is a risk worth taking. It is only through democratic debate that citizens clarify the issues that face us and decide what we deem to be the political truth (regardless of what the experts instruct us to believe). But it is much more than that.

In a real democracy, what matters is not just the outcome but the way that citizens participate in a debate about the future. That participation is what enables us to take meaningful moral responsibility for our actions, and develop the spirit of freedom that keeps public life healthy. The problems really begin, not when democracy goes 'too far', but when the people are taken for granted or ignored.

In his two volumes of *Democracy in America*, Tocqueville observed two sides of what he described as the problem of 'democratic fatalism' – the belief that history was on the side of democracy. This could lead to 'recklessness as well as resignation'. In volume 1 Tocqueville discusses the danger of 'the tyranny of the majority', the 'moments when democracies run wild' in potential race riots or outbursts of war fever. In volume 2, he describes the 'mild despotism of public opinion', which, notes the historian David Runciman, 'is more insidious and makes people reluctant to challenge conventional ideas. As well as running wild, democracies can stagnate.'[41]

The first of those problems is most often cited as the danger facing our democracies today, with the outpouring of fears about the 'tyranny of the majority' supposedly inflamed by racist and other 'populist' demagogues. Yet the second of Tocqueville's symptoms surely more accurately captures our current situation: a formal democracy not running wild but stagnating under the 'mild despotism of public opinion' and an insidious culture of conformism.

To stir our spirits and seek to break out of that stagnation, we need the widest no-holds-or-opinions-barred democratic debate about everything from Brexit to bans on 'extremism', the economy to energy policy and back. In short we need more living, dynamic democracy, and less brain-deadening conformism and convention.

The prevailing attitude of 'I'm a democrat, but …' is on a par with that other reasonable-sounding motto of our times, 'I believe in free speech, but …' In reality, however, you are either for free speech and democracy, or you are not. Anything else is a pretence, a fraud on free speech or a mockery of democracy, akin to protesting that you believe in God but only when He does what you want.

Democracy is about choice. That does not mean, however, that you get to choose which bits of democracy you tolerate, or cherry-pick the votes you are prepared to live with. Qualifying your support for democracy in practice puts a big cross through the principle of voting and democratic representation.

As with free speech, our support for democracy only really matters when we get to the 'but' part. It is always easy to support free speech for those who agree with you. The acid test, however, as a US Supreme Court judge put it nearly eighty years ago, is not backing freedom for ideas we like but 'freedom for the thought that we hate'. Or as George Orwell has it, in his proposed 1945 preface to *Animal Farm*, entitled 'Freedom of the Press': 'If liberty means anything at all, it means the right to tell people what they do not want to hear.'[42]

In similar vein, we might say that the acid test for democracy is accepting the freedom of others to stand for ideas that we hate, and the right to vote for what we do not want to happen. That does not mean anybody has to passively accept the outcome; the right to argue and protest against a referendum result or an elected government is a fundamental democratic freedom. It must mean, however, not trying to bypass or subvert democracy through the backdoor, via the courts or other unaccountable bodies.

Just after the UK referendum, a US writer at *Rolling Stone* magazine highlighted the fashion for American liberals and intellectuals to attack Brexit and Trump as problems of 'too much democracy'. Matt Taibbi linked these trends to the new assaults on free speech on US university campuses, often led by student activists, and to Trump's own demands for a crackdown on immigration. 'Democracy appears to have become so denuded and corrupted in America,' wrote Taibbi, 'that a generation of people has grown up without any faith in its principles.' We might only add that the loss of faith in democracy is not confined to America.

As he concluded, 'If you think there's ever such a thing as "too much democracy", you probably never believed in it in the first place. And even low-information voters can sense it.'[43]

Let us remind the new critics of democracy of the meaning of the word. It does not mean technocracy, bureaucracy, or krytocracy (rule by judges). People did not fight and even die down the centuries for the right to be ruled over by a Clerisy of professional politicians and experts. Democracy must mean in essence the people exercising power. By that standard the problem is not 'too much' democracy, but that we have never had enough of it. As the British historian E. H. Carr suggested back in 1951, when the West was still celebrating the triumph of its democracy over fascism and the stand of the 'free world' against Communism: to speak about defending democracy as if it were something we had long enjoyed was 'self-deception and sham'. Some states might be more democratic than others, Carr conceded; but none was very democratic, to judge by 'the question where power resides and how it is exercised'. 'Mass democracy', he concluded, remained 'largely uncharted territory', and 'we should have a far more convincing slogan, if we spoke of the need not to defend democracy, but to create it.'[44]

It might help to remind ourselves of how hard-fought has been the historic struggle for democratic liberties. The next chapter outlines how democracy has always been a dirty word for those who believe that they know what's best for the rest.

# 3

# A short history of
# anti-democracy

Conventional wisdom has it that the history of Western politics has been a forward march of democracy down the centuries, leading to the inevitable triumph of the democratic system we enjoy today. But some versions of history really do need rewriting.

As Anthony Arblaster, author of *Democracy*, observed at the turn of the twenty-first century, 'It might be better to write a history of the opposition to democracy. It would make a more substantial volume than the history of democracy itself, and it would remind us that democracy, where and in so far as it has been achieved, has been achieved against the odds, and against fierce resistance.'[1]

We in the Western world like to congratulate ourselves on our unique history of democracy. In the US, the democratic tradition can be traced back to the American Declaration of Independence of 1776, which stated with characteristic certainty that 'all men are created equal, that they are endowed by their Creator with certain unalienable Rights, and that among these are Life, Liberty and the pursuit of Happiness'.

In Britain we boast (in our modest, British way) of a longer democratic legacy stretching back 800 years, to the sealing of

Magna Carta in 1215. The charter through which a group of English barons and bishops forced King John to grant certain rights including a fair trial to 'free men', and to cede that the Crown could not impose taxes without obtaining the consent of the nation through a common council of lords, is often hailed as the foundation stone of justice and parliamentary democracy.

And the origins of the Western democratic tradition can be found buried deeper still, some 2,500 years ago in the golden age of ancient Greek civilisation and Athenian democracy, in the fifth and fourth centuries BC.

The danger this view of history poses to the present is that, as Arblaster observed, 'Complacency is nourished, and resistance to democracy is persistently under-estimated and under-played.' Before giving ourselves yet another democratic pat on our collective back, a quick look behind those historical headlines might suggest a slightly different version of events.

It can remind us that there is nothing 'inevitable' about democracy; that it should never be taken for granted; that it has had to be fought for and defended over and again through modern times; and that even today, when every leading voice in the West is nominally a democrat, the real meaning of democracy for the masses remains unresolved and up for grabs.

Much of Western history looks more like a long story of anti-democratic thought and political action from society's elites. It is a tale of those with power using it to restrict the spread of democracy and contain the aspirations of the masses for more freedom of choice and control over their lives.

The forces opposed to real democracy have often changed with the times, from the oligarchs of the ancient world through the aristocrats of yore to the more liberal elitists of the modern era. Yet a short history of anti-democracy reveals a remarkable underlying similarity in the arguments they use down the centuries, always expressed in the language of the day.

From the great philosopher Plato to the less elevated elitist thinkers of today, opponents have argued against extending democratic control in society on five closely connected grounds:

1. People are ignorant at best, or simply idiots. That most people are too uneducated to understand what is at stake in important political debates, and big issues are too complex for them to grasp. It is unfair and dangerous to expect such small minds to make major decisions, at least without being tutored by the experts.

2. People are irrational. That many people are led by their emotions rather than intellect, driven by passions rather than reason, more likely to demand a war than support world peace. Such childish citizens are not to be trusted with adult decisions and a role in grown-up government.

3. People are gullible. That most people are as soft and pliable as putty, easily manipulated by the promises and lies of political demagogues, the mass media and corporate advertisers. Democracy really means giving power to these few Pavlovian puppet-masters.

4. People are selfish. That most people are inhumane, and have so little feeling for their fellow men and women that they will generally act malevolently to get whatever they want. Handing democratic control to them is the same as empowering a hate-filled lynch-mob.

5. People don't know what's good for them. That the uneducated, irrational, emotional, malicious masses have no idea of their real interests. Give them the right to vote and they will support the wrong things. Better to leave it to enlightened elites who know what's best for the rest.

In different forms these arguments against giving greater power to the mass of the people make frequent reappearances on the stage of history.

In practice democracy has had a pretty short life. It certainly existed in ancient Athens, albeit not in a form we would recognise today. Democracy then disappeared from the political map for more than 2,000 years, the D-word itself barely even recognised or spoken of, in a long, dark age when the only 'right' recognised was the divine right of kings to rule.

Every time there has been some apparent advance towards democracy, that movement has been met with powerful attempts to contain it. Often the reversal was abrupt; far from lasting for 800 years, for example, the original Magna Carta of June 1215 barely lasted ten weeks before it was declared null and void by Pope Innocent III, outraged at the idea that any man could try to limit the power of a king anointed by God. More than 600 years later, the revolution in France of 1848 established the Second Republic and universal male suffrage. That advance lasted barely two years before the horrified forces of French conservatism overturned universal male suffrage, soon followed by a *coup d'état* and the establishment of the Second French Empire under Napoleon III.

Only in the past 200-odd years has democracy come to be accepted as a legitimate political idea or form of government at all – and even then, exactly what we might mean by democracy has been fiercely contested with words and other weapons. In the opinion of one historian, 'Crudely speaking, up to the 18th century everyone had a clear idea of what democracy was and hardly anyone was in favour of it. Now that position is reversed. Everyone is in favour of it but no-one has a clear idea any longer what it is.'[2]

## Athens: philosophers against the people

To start with it is worthwhile to take a short glance back at the Athenian experience, as it foreshadows the themes of arguments around democracy ever since.

Democracy is widely considered the watchword of classical Athenian civilisation. Yet think of the best-known ancient Greek thinkers – the philosophers Socrates, Plato and Aristotle. The first two were fiercely anti-democratic, and the third was no lover of democracy. Indeed it has been said that, to judge by the relatively few surviving texts, Western political thought began in ancient Greece with the express aim of criticising and curtailing Athenian democracy. As J. S. McClelland, author of *The Crowd and the Mob*, has it, 'It could almost be said that political theorising was *invented* to show that democracy, the rule of men by themselves, necessarily turns into rule by the mob [author's emphasis].'[3]

If we think of our society's elites as being somewhat wary of democracy, in ancient Athens they were truly terrified by it. The Athenian system was based on the principle of full-blooded direct democracy, not the comparatively pallid representative system we are familiar with today.

The male citizens of Athens had a direct say and vote on their city-state's laws and rules, through the Assembly and the people's juries. Most public posts were sortitive rather than elective, filled at random by a lottery of the citizens; the jurors who sat in judgement of key cases and defined Athenian law came from a panel of 6,000 citizens, chosen by lot each year. This direct exercise of people power became even more popular once Pericles, the greatest of Athens' democratic leaders, shifted more powers from the elite Council to the people's courts and introduced payment for jurors, enabling lower-class citizens to take part more readily in decision-making.

The overall effect, according to Paul Cartledge's recent history *Democracy: A Life*, was that what most ancient Athenians 'understood both practically and symbolically by *demokratia* was something much more like Lenin's revolutionary Bolshevik slogan, "the dictatorship of the proletariat"'. The system we call democracy, he adds, would be viewed 'dismissively or contemptuously' by a committed Athenian democrat as 'at best – disguised oligarchy', the rule of a few.[4] (The extent of democracy under this peculiar dictatorship of the proles was obviously limited by the fact that it excluded all women and the many slaves of ancient Athens.)

*Demokratia* in Athens was not just a technical means of choosing leaders, but meant the direct exercise of political power. Above all, it was a living system of participation. Its most famous exponent was Pericles, whose democratic state-building led to the majestic Parthenon atop the Acropolis in Athens and the flourishing of art and theatre, notably in the tragedies of his friend Sophocles.

Pericles' funeral oration for the dead of the first Peloponnesian war, quoted in chapter 2, is widely seen as the finest statement of the meaning of Athenian democracy. It stressed, says Jennifer Tolbert Roberts in *Athens on Trial*, 'the creative, dynamic power of democracy to unite men of all classes in active participation in the government'. More than simply offering formal legal equality, it meant 'rather the presence of a positive and vital force drawing each (male) heart and mind into both deliberation and action, a virtue lacking under other constitutions.'[5]

This popular participation not only helped make democratic Athens more virtuous than other systems, but more culturally vibrant and militarily victorious too; as Frank Furedi, author of *Authority*, says, 'Athenian society worked most of the time because its people identified with their city'. Authority was invested not in an elite but 'in the people and the opinions they expressed through the Athenian assembly and other public venues'. The authority of 'public opinion and its democratic culture' was boosted by historic events such as the defeat of the Persians at the battle of Salamis, where it was 'the poor sailors of the Athenian navy' who won the glory, rather than 'the heroic upper-class warriors' of classical mythology.[6]

Another funeral oration by a champion of Athenian democracy proclaimed it 'the way of wild beasts to be held subject to one another by force, but the duty of men to delimit justice by law, to convince by reason, and to serve these two in action by submitting to the sovereignty of law and the instruction of reason'. As the historian Jennifer Tolbert Roberts concludes, 'To be undemocratic, in other words, was to be – literally – inhuman.'[7]

Unsurprisingly, the Athenian elites were never comfortable living under such a system of dynamic democratic power. The '*kratos*' root of *demokratia* could mean not only control, but grip or grasp. The minority of wealthy oligarchs feared that, unless kept in check, the system would allow the grasping lower orders to get a grip on their assets.

Plato was an intellectual authoritarian bitterly critical of the fundamentals of democracy. In the *Gorgias* (*c.* 380 BC), Plato has the master philosopher Socrates (who never wrote anything down himself, so we only have Plato's word for it) branding the most respected political leaders as little better than pastry chefs, since all they did was pander to the base appetites of the drooling masses. Plato gave full vent to his spleen in his great work *The Republic*, where he lays into democracy as the lowest form of government. Why, in a true democracy, scoffs Plato's Socrates, even the donkeys would assume airs! While democrats viewed democracy as the highest expression of their shared humanity, elitist philosophers equated the lower class of citizens with mere beasts of burden.

Plato thus held, says Cartledge, 'in a sort of anticipation of George Orwell's "Some animals are more equal than others", that some citizens are as it were more equal than others'.[8] Since a few citizens were evidently superior in terms of everything from their birth and looks to education and intellect, it was only right and fair that they should enjoy a bigger share of power than the rest. Rather than the democratic tyranny of the masses, he favoured the rule of the few – preferably rule by a few wise Platonic philosophers, the only ones deemed truly capable of good government.

Others went further still in spelling out their prejudices and their determination to put them into practice; Aristotle recorded that the extreme oligarchs in some Greek cities had taken a secret oath that 'I will be evil-minded to the *demos* and will plot whatever evil I can against them'.[9]

The anti-democratic philosophers denounced the political leaders of their age as 'demagogues', an insult which has come down the ages to us via historians, meaning rabble-rousers who only lead the gullible mob astray in pursuit of evil ends. Yet in the original Greek, '*demagogas*' meant merely 'leader of the *demos*', the people. We have apparently accepted the oligarchs' bigoted view of that as an inherently dangerous role.

## The revolt of the 'mad men'

When the popular system of Athenian democracy was finally overthrown – not by the oligarchs within, but by Philip II of Macedonia in 338 BC – it was quickly all but erased from history. Not only was the system forgotten in the long centuries of empire, monarchy and caliphate, but the D-word itself effectively disappeared from the languages of the West.

It would take 2,000 years before democracy could mean something again. The European age of Renaissance and Enlightenment brought ideas of autonomy, liberty and equality to life, as people sought more of a say in the dawning modern age. The growing belief in the autonomy of the individual, freed from the bonds of feudalism, was accompanied by fresh ideas and demands for a greater say in the shaping of society.

However, even as the Enlightenment dawned, many thinkers were uncomfortable with the idea of democracy, and it was often those reacting against the new demands for democratic change and political experiments who left the strongest impression on history. The French writer Jean Bodin published his popular *Six Books of the Commonwealth* in 1576 (English edition 1606) to attack 'the evils of popular government' as practised in ancient Athens: 'How can a multitude, that is to say, a Beast with many heads, without jugement, or reason, give any good councel? To aske councel of a Multitude (as they did in old times in Popular Commonweals) is to seek for wisdome of a mad man.'[10]

As late as 1609 King James I of England still felt able to declare that 'Kings are justly called Gods, for that they exercise a name or resemblance of divine power upon Earth.'[11] Despite James's divine self-image, the democratic demand to call such kingly Gods to account before the people was already bubbling under the surface of British society.

It burst forth beneath his son, King Charles I, in the tumult of the English revolution (1642–51) – often called the civil war, sometimes by those who still cannot contemplate the notion of supposedly stable, conservative England staging bloody revolution and beheading a king, almost 150 years before the revolting French got the idea.

The revolt by parliamentary forces against absolute monarchy was initially led by conservative men and aristocrats. Yet the English revolution soon mobilised the masses who wanted more far-reaching changes to the system of government.

This was when the modern world heard arguably the first proto-democratic voices of the people: the Levellers, the radical movement for liberty led by the likes of 'Freeborn' John Lilburne and Edward Sexby. Lilburne was frequently imprisoned and flogged by the Crown for writing radical, intemperate pamphlets such as *The Freeman's Freedom Vindicated*, which declared it 'unnatural, irrational, sinful, wicked, unjust, devilish and tyrannical' for 'any man whatsoever' to assume 'a power, authority and jurisdiction to rule, govern or reign over any sort of men in the world without their free consent'.[12]

While they supported parliament against the king, the Levellers would not settle for being ruled by the existing aristocratic and unrepresentative crop of parliamentarians, either. In the 1640s the majority of MPs were from small rotten boroughs where their seats were bought and sold by bribery, patronage and royal appointment. Even where there were elections, only a tiny percentage of property-owning adult males and no women were entitled to vote. The result, observes Paul Foot, author of *The Vote*, was that 'the representative element of Parliament was next to nothing'.[13]

The Levellers demanded radical reforms. In 1646, Lilburne's pamphlet *London's Liberty in Chains* listed proposals to shake up the system and break the lords' and landowners' monopoly on power: annual parliaments; the publication of parliamentary proceedings and an end to secrecy; payment of MPs; equal constituencies and the abolition of rotten boroughs. These ideas provided the spark for a remarkable experiment in direct democracy.

The first true directly elected democratic body in England was formed, not in the aristocratic atmosphere of the Westminster parliament or in some local parish council, but within the rank and file of Oliver Cromwell's New Model Army. In 1647, as discontent grew in the ranks, the Levellers' pamphlet *The Case for the Army* demanded representative government. 'This power of Commons in Parliament is the thing against which the King hath contended, and the people have defended with their lives, and therefore ought now be demanded as the price of their blood'.[14] They demanded votes for all, except the 'delinquents' who fought for the king.

But the army was not waiting to be granted the vote from on high. They created the Army Council, made up of agents or agitators directly elected by common soldiers, and marched on London to make their demands heard.

The parliamentary elite reacted to this democratic rebellion with as much horror as any king could have mustered. Leveller petitions were ordered to be burnt by the hangman, petitioners arrested at Westminster and imprisoned. But the power of the rank and file forced the elite to contend with the challenge. In October and November 1647 the Putney Debates, held in a south London church, brought Cromwell and other commanders together with Levellers and army representatives to discuss what sort of system should replace absolute monarchy – an unprecedented sort of 'parliament'.

The Putney Debates are rightly remembered not least for the

speech made by Leveller spokesman Colonel Thomas Rainsborough MP, who dared to suggest that the poor should be asked to 'consent' to the system of government under which they lived, and declared that the 'poorest man in England is not at all bound in a strict sense to that government that he hath not had a voice to put himself under'.[15] Although Rainsborough and his fellow Levellers never uttered the D-word, the response to the debates introduced the idea of democracy into English political discourse for the first time.

In the end Cromwell, who looked down with disdain upon such upstart notions, won the struggle with the Army Council and the war with the king and became Lord Protector of the Commonwealth. The Levellers were crushed and forgotten, their very existence 'levelled' on the field of history until being revived two centuries later.

During the revolutionary era the demand for greater liberty in speech and politics was also taken up by important figures such as the poet John Milton. As often, however, the highbrow anti-democratic voices responded all the louder to the merest suggestion of giving more freedom to the people. In his epic work *Leviathan* (1651), the philosopher Thomas Hobbes decried the fact that educated men such as Milton had been taught to read the classical works of ancient Greece and Rome, which radical nonsense had given them 'under a false show of liberty, a habit of favouring tumults, and of licentious controlling the actions of their sovereigns ... with the effusion of so much blood'.[16] For the anti-democrats, it seems, the uneducated masses were not the only problem – over-educated men filling their heads with 'licentious' notions were even more dangerous.

As quickly as 1660 the British monarchy was restored under Charles II. (Most of the main players in the Putney Debates were already dead, the Levellers Lilburne and Sexby having both perished in the Tower of London along with the liberties they fought for.) Then in 1688 the Protestant William Prince of Orange was invited in from the Netherlands to overthrow the Stuart dynasty and their delusions of divinity for good. William and his English wife, Queen Mary, oversaw the 'Glorious Revolution' – a top-down reform of government that gave more powers to parliament but was in fact neither particularly glorious nor very revolutionary. The vast majority of British people still had no vote or voice in deciding the way they were governed.

It was in this age of limited change that John Locke wrote his *Two Treatises of Civil Government* (1690). Locke has often been hailed as the 'intellectual godfather of liberal parliamentary democracy' on both sides of the Atlantic. Yet the popular consent for government he proposed was 'still passive, not active', his suggested political contract a far cry from democracy and 'loaded in favour of the State rather than the People'.[17]

Across the sea in continental Europe, meanwhile, more forward-looking thought was lighting the torch for democracy. Baruch Spinoza, the Great Dutchman of the Enlightenment, was a key figure in laying the intellectual and moral foundations for many of our modern ideas of freedom.

It was Spinoza who set the standard for freedom of speech which, almost 350 years later, Western societies still struggle to meet: 'In a free state, every man may think what he likes and say what he thinks'. And it was Spinoza who, in his controversial 1670 work *Tractatus Theologico-Politicus* (*Theologico-Political Treatise*), spelt out his belief that the best form of a free state was a democracy, by which he meant representative government – one of the very first explicit endorsements of democracy in the modern age, filled with sufficiently dangerous ideas to prompt Spinoza to publish it anonymously.

'In a democracy,' wrote Spinoza, 'irrational commands are still less to be feared: for it is almost impossible that a majority of a people, especially if it be a large one, should agree in an irrational design: and, moreover, the basis and aim of a democracy is to avoid the desires as irrational, and to bring men as far as possible under the control of reason, so that they may live in peace and harmony'.

Spinoza concludes: 'I believe it [democracy] to be of all forms of government the most natural, and the most consonant with individual liberty. In it no one transfers his natural right so absolutely that he has no further voice in affairs, he only hands it over to the majority of a society, whereof he is a unit. Thus all men remain as they were in the state of nature, equals.' Spinoza's elevation of democracy founded on reason as the best route to 'freedom in a state', alongside his criticism of monarchy and critique of established religion, were sufficient to get the *Treatise* banned, even in secular and relatively enlightened Holland.[18]

As the modern age dawned in the eighteenth century, political change was in the air. Yet in England, the supposed home of freedom and democracy, even the reformers were keen to keep the untrustworthy masses at arm's length. Henry Fox, father of the leading Whig reformer Charles James Fox, might have been considered a liberal voice in the 1700s, but such liberalism had strict limits when it came to extending the right to vote. 'When we talk of the people with regards to elections,' declared Fox, giving voice to the thoughts of many reformers, 'we ought to think only of those of the better sort, without comprehending the mob or mere dregs of the people'.[19]

Denied any political representation, the mass of British people could only express their preferences through that pre-democratic form of popular expression, the riot – as when

50,000 Londoners besieged parliament in 1771 in defence of the radical journalist, campaigner, MP and rogue John Wilkes, crying, 'Wilkes and Liberty!' and almost lynching the prime minister, Lord North. The Wilkes riots and similar outbursts of popular protest only served to make the elites even more fearful of granting the majority more power.

## Revolution – but not 'mob rule'

After its 2,000-year slumber, democracy was to be truly reawakened in explosive style by the American and French revolutions of the late eighteenth century. Yet even amidst those tumults, fear and contempt towards 'the mob or mere dregs of the people' was much in evidence.

In 1776 the Founding Fathers of the United States had declared the equality of all men in order to mobilise and justify their revolutionary war of independence against the British colonialists. Yet most of these future US presidents were not really committed democrats, instead displaying an unbridled passion for a strictly bridled form of democracy. They favoured the order of ancient republican Rome over the turbulence of democratic Athens, and accordingly named their government institutions – Senate, Capitol – after the Roman bodies.

In his contributions to *The Federalist Papers* James Madison, later the fourth US president, advanced some of the key arguments over the shape of the new American government. Madison was keen to draw a line between the American republic and those 'turbulent democracies of Ancient Greece and modern Italy'. The problem with popular assemblies, he insisted, was that the passion of the irrational masses would inevitably drown out the reason of wise men. Even if every Athenian citizen had been a Socrates, Madison maintained, 'every Athenian assembly would still have been a mob'.[20]

This was why Madison argued for an upper house, the Senate, to be less democratic and empowered to protect people from their own worst instincts, whenever they were 'stimulated by some irregular passion' 'or misled by the artful misrepresentations of interested men' to 'call for measures which they themselves will afterwards be the most ready to lament and condemn'. The powerful Supreme Court and the electoral college would be other arms of the system of constitutional 'checks and balances' to keep the 'irregular passions' of the populace in check.

Madison's case against potential mob rule won the day, and the US Senate was established as a conservative bulwark against too much democracy in the lower house of Congress. (The allocation of two Senate seats to each US state regardless of its population, for example, being intended to help the smaller rural and generally more conservative states counter the popular influence of the big urban centres.) It seems that those anti-democrats who have argued for the wise unelected men and women of the House of Lords to protect the British people from their own irregular passions by reversing the Brexit referendum vote have some auspicious forebears.

Others among the Founding Fathers were even more adamant that America would be a republic, but not a popular democracy. John Adams, the second US president, was horrified by any notion of majority rule, which he saw as a mortal threat to the all-important property rights of America's new ruling class. Adams predicted it would mean 'The idle, the vicious, the intemperate would rush into the utmost extravagance of debauchery' at the expense of the propertied minority, with the result that 'anarchy and tyranny commence'.[21] It was to allay these fears that the Founding Fathers refused to countenance any idea of universal white male suffrage, never mind votes for white women or, of course, for black slaves (most of them were slave-owners).

Looking back at the revolution and its legacy today, it has become quite fashionable in some quarters to be dismissive of American democracy because it was founded on slavery. The slave issue certainly reveals the limits of the Founding Fathers' feel for democracy. Yet it is also possible to draw a more positive conclusion: that the creation of the American republic and the spirit of democratic liberty it was infused with laid the basis for the freedom struggle that would abolish slavery in the nineteenth century. Democracy is an unpredictable and sometimes uncontrollable force that, once unleashed, can confound the moderate ambitions of those who opened the door for it.

One key figure in the American revolution who argued for a much wider extension of democracy was the English radical Tom Paine, author of *Common Sense* and *The Rights of Man*. One of Paine's arguments for democracy was that it would enable the abolition of the 'savage practice' of slavery – a shocking idea, even in revolutionary America.[22] His case proved too radical for the Founding Fathers of the US republic, and Paine departed for France, where he also became a leading light in the second great popular revolution of the late eighteenth century.

The French revolution that began in 1789 is often remembered and taught in the English-speaking world as a dark time of violent turmoil and terror, symbolised by Madame Guillotine. Yet in overthrowing (and beheading) the most powerful absolute monarchy in Europe, and establishing the principle of representative government via the National Assembly, the French revolution and its slogan '*Liberté, Egalité, Fraternité*' lit a beacon for democracy in Europe that still burns. The Declaration of the Rights of Man and of the Citizen abolished feudal bonds and privileges, established the principle of representative government and enshrined equal rights including 'liberty, property, security and resistance to oppression' for all French citizens. The revolutionary era ended in the Great Terror followed by the military dictatorship and then empire of Napoleon Bonaparte. But the Enlightenment ideas of liberty and rights that the revolution initially embodied terrified the autocratic rulers of the rest of Europe, who not only tried to overthrow the French revolution by force but also set about reforming their own systems of government to try to contain the spread of the spirit of popular revolt.

It was around the French revolution that Tom Paine was to become involved in one of the defining debates of Western (anti-)democratic politics. His opponent was Edmund Burke, now considered the founding father of modern conservatism. Burke was horrified by the mass mobilisations and terror unleashed by the French revolution. He recoiled not only from the violent turmoil, but also from the underlying Enlightenment principles of democratic consent and individual choice. Burke took the view that people don't always understand what's good for them. The people need the authority of tradition – what has been called 'the democracy of the dead' – guided by a wise ruler, to move them in the right direction.

Paine, by contrast, believed in the power of reason and the rational individual. Unlike Burke and most of his contemporaries, Paine was a radical who sided with the revolutionaries because he upheld the liberal Enlightenment values of individual rights and the principle of democratic consent. Paine wanted to free people from the constraints imposed by the past and traditional authority, to put the living over the dead, and enable them to remake their societies; as he put it in typically forward-looking fashion, 'We have it in our power to begin the *world* over again.' These ideas seemed threatening even to those among the political elites of the eighteenth and nineteenth centuries who considered themselves liberal-minded, yet doubted the power of reason over the mob and felt uneasy about recognising the right of individuals to act as they saw fit.

Burke saw the need for political reform to contain the threat of further revolutionary upheavals in Britain and elsewhere. But he believed that the political class, even if it must submit to being elected, must remain above the masses, ruling as the elite saw fit in what it considered to be society's best interests. Burke wrote that those men lacking 'honour', who held such lowly occupations as a hairdresser or a candle-maker, should not 'suffer oppression by the state' on account of their 'servile employments'; but he was adamant that 'the state suffers oppression, if such as they, either individually or collectively, are permitted to rule'.[23] He upheld the importance of monarchy as the mainstay of traditional British authority. Paine took the democratic view that politicians should represent those who elected them, carrying out the popular will, free from the stifling weight of monarchy and tradition.

As the Western elites took fright at the forces unleashed by the French revolution, it was Burke who ultimately won out in that contest. Our societies' view of what democracy means has unfortunately been shaped by that outcome ever since.

## Victorian liberals vs the 'ignorant brutish masses'

The nineteenth century looks like a period of advance for democracy, as people fought for their rights and electoral reforms extended the franchise to more and more voters in both Britain and America. Tocqueville's classic *Democracy in America*, volume 1 first published in 1835, volume 2 in 1840, is notable as one of the earliest modern works ever to use the

D-word in a positive sense, rather than as a mob-handed bogeyman to be feared.

In Britain, meanwhile, the movement for extending the franchise gathered steam, winning a series of electoral reforms that extended the vote to a larger proportion of men. Between 1838 and 1858 the Chartists, a working-class movement for political reform, fought for the six radical demands of the People's Charter, which millions of people signed up to support:

1. A vote for every man twenty-one years of age, of sound mind, and not undergoing punishment for a crime.

2. The secret ballot to protect the elector in the exercise of his vote.

3. No property qualification for Members of Parliament in order to allow the constituencies to return the man of their choice.

4. Payment of Members [of Parliament], enabling tradesmen, working men, or other persons of modest means to leave or interrupt their livelihood to attend to the interests of the nation.

5. Equal constituencies, securing the same amount of representation for the same number of electors, instead of allowing less populous constituencies to have as much or more weight than larger ones.

6. Annual Parliamentary elections, thus presenting the most effectual check to bribery and intimidation, since no purse could buy a constituency under a system of universal manhood suffrage in each twelve-month period.

Drawing on the tradition of the Levellers, these demands for real reform, which still sound radical today, made the ruling elites even more uneasy about the very idea of democracy. They sought to contain it by redefining and sanitising the concept into something that would not threaten their control.

Thus each electoral reform in Britain – in 1832, 1867 and 1884 – extended the male franchise further, but always within limits. Property qualifications for voting were reduced, but not abolished. This was important because ownership of property was seen as a stake in the status quo, a sign not only of wealth but of respectability and moderation. The propertyless masses, by contrast, were no more to be trusted than the *plebs* of the Roman mob. And all women were still denied the vote, whether the members of that untrustworthy sex owned property or not.

Contrary to another myth of history, none of these electoral reforms was granted out of the goodness of the government's heart. Each stumbling step forward for democracy was only achieved because mass movements fought for it. Always underpinning this reticence about extending democracy was the fear of the masses as an irrational, easily swayed mob. In this elitist view the growing popular protests for democratic reform were themselves evidence of the dangers of democracy. Thus when a mass demonstration for representation at St Peter's Field in Manchester – a booming industrial city that still had no MPs – ended in the Peterloo Massacre of 1819 as troops charged the crowd, the first response of the government was not to grant reforms but to enforce new punitive measures against demonstrations and the free press.

The people's democratic aspirations, however, were not so easily to be repressed. After the Peterloo Massacre the radical poet Shelley penned 'The Mask of Anarchy', capturing the spirit of the age:

Rise, like Lions after slumber
In unvanquishable number –
Shake your chains to earth like dew
Which in sleep had fallen on you –
Ye are many – they are few.

As the protesters continued to rise and shake their chains in increasingly violent style, the British authorities finally moved to pass a Great Reform Bill – not to meet the demand for democracy, but to contain it. The Bill would abolish rotten boroughs and introduce MPs for the metropolitan centres, whilst extending the right to vote – but only to a larger section of property-owning males.

Even this proved too much for the House of Lords. In July 1831 the Duke of Wellington, hero of Waterloo, wrote to a fellow noble spitting venom at how 'the mob, the Radicals, the dissenters from the Church of all religious persuasions hail the Bill as the commencement of a new era of destruction and plunder'.[24]

The Lords duly voted the Bill down in October 1831. The popular response put Shelley's poetic call into practice. Riots broke out across the country, several cities were soon in flames, and in Derby the massive castle of the Duke of Newcastle, a forthright opponent of reform, was burned to the ground. Further west in Bristol the anti-Reform Bill MP Sir Charles Wetherell had declared in horror that he could not lower himself to seek the support of the proposed new electors, who might live in property valued at only £10 a year, as it would mean touting for votes 'in the lazaretto' – an isolation hospital for those suffering leprosy or the plague. In response, Wetherell's disenfranchised constituents staged an armed revolt, taking over the city for a week and burning down Bristol's four prisons along with other major buildings. The sniffy MP himself, who had declared he would 'uphold the King's peace' against the mob, was 'forced', as Paul Foot recorded with some relish, 'to flee over the rooftops in his underclothes in the middle of the night'.[25] As the panic spread from Derby and Bristol to London, parliament finally and begrudgingly conceded the Great Reform Act in 1832.

It is striking how many fought and rioted for that first Reform Act who were themselves not even to be granted the vote by its limited extension of the franchise. But the spirit of democratic change was in the English air, and it is a hard spirit to contain once a whiff of it escapes.

In continental Europe, meanwhile, the working classes made their first appearances on the political stage in the series of democratic revolutions that shook crowns and governments in 1848. While Karl Marx and Frederick Engels wrote *The Communist Manifesto*, in France the second revolution led to universal male suffrage, granting the vote to 9 million more men. But as soon as the tide of popular revolution ebbed again, the French elites moved to take back that concession as early as 1850.

The riotous debate in the assembly was marked by Victor Hugo's stirring defence of universal suffrage and elections as 'one day in the year when the most invisible citizen, the social atom is involved in the vast life of the entire country, a day when the weakest feels in himself the greatness of national sovereignty, where the most humble feels in himself the soul of the country'. This, Hugo continued, 'is because he dares to use his vote to his fancy. These people seem to have the audacity to imagine that they are free. And apparently another strange idea comes into their head that they are sovereign.'

Hugo launched a coruscating attack on the ignorance and animosity which France's parliamentarians displayed towards the people. 'Because they have the insolence to give an opinion in this peaceful form of the ballot and not to bow down altogether at your feet, then you are indignant, you get angry, you cry: we will punish you, people! We will punish you, people! You'll have to deal with us, people! It is your ignorance of the current country; the animosity that you feel for it and that it feels for you.'[26] Hugo lost the debate, universal suffrage was overturned, a coup established the Second French Empire and he fled France. Man was once more temporarily reduced to a 'social atom'.

In republican America things went further down the democratic path – white male suffrage was almost universal by 1856 and women were even granted the vote in Wyoming in 1869. Elsewhere in America, however, women were still barred from democracy, along with black people, whether enslaved or formally freed.

Indeed, even in America the extent to which some were already seeking to redefine 'democracy' and empty it of meaning was illustrated by the role of the misnamed Democratic Party as the champion of slavery in the era of civil war. Before that war began, future abolitionist president Abraham Lincoln took part in the famous Illinois election debates of 1858 with Democratic Party Senator Stephen A. Douglas. These exchanges were notable for the way Douglas sought to twist the meaning of democracy effectively to defend legalised slavery. He claimed that the principle of 'popular sovereignty' meant the local white males-only electorate should have the democratic right to vote for the slave system. Since slaves could not vote, they could not count in a democracy. Lincoln dismissed this perversion of 'popular sovereignty' as 'a living, creeping lie'.[27]

The triumph of President Lincoln's Union forces in the subsequent American civil war led in 1870 to the 15th Amendment to the US Constitution, granting the vote to all adult males of whatever race, including former slaves. It was not too long, however, before the forces of anti-democracy had regrouped; by the turn of the century Southern states would be passing 'Jim Crow' laws, using poll taxes, literacy tests and other rigged rules to disenfranchise black voters. This stain on US democracy would last until the civil rights era of the 1960s.

## Liberals against the masses

Back in Victorian Britain, it was striking then as now the extent to which, while working-class movements such as the Chartists fought for democracy, among the middle classes even the pro-reform liberal voices of the age feared the democratic unleashing of the 'dregs of the people' below.

The Chartists' demands for the vote were met with familiar-sounding conservative objections, insisting that 'universal suffrage could never be of great practical utility, unless bestowed upon a well-informed, intelligent public'. In response leading Chartists insisted that working people were actually better qualified to decide on the lives of the majority, since they were 'aware of the particular hardships that oppress them, and men who do not feel these hardships seldom are. The poor man feels where the shoe pinches him, and from an intimate knowledge of his affairs, can easily determine how any measure would affect his interest, and better his condition.'[28] These working-class radicals wanted the vote, not simply as an end in itself, but as a means to better the conditions of the masses.

The middle-class horror at such radical ideas was not confined to conservatives. Liberal society reacted with fear at the prospect of wider democracy, too. Thus in the midst of the agitation preceding the 1867 Reform Act, including a major riot in which every iron railing around London's Hyde Park was torn up and the police were pelted with missiles, the liberal author George Eliot (real name Mary Ann Evans) published her novel *Felix Holt, the Radical*, about the unrest that accompanied the first Reform Act of 1832 in a Midlands town. It is essentially a warning about the dangers of unleashing democracy when the uneducated masses can be misled by demagogues and scheming political agents seeking to buy their vote with beer. Felix Holt, the alleged 'radical', is actually opposed to giving the vote to the nearby mining community, believing that 'extension of suffrage … can never mean anything for [working men] but extension of boozing'. Instead he wants a programme of education and moral improvement to convert the masses into respectable citizens worthy of society. Such were what passed for 'radical' views in the liberal England of the mid-nineteenth century.[29]

More telling yet was the attitude towards democracy espoused by John Stuart Mill MP, the greatest British philosopher of the Victorian age. To those of us who fight to defend freedom of speech against all-comers today, J. S. Mill is a hero of history, a giant on whose shoulders we try to stand, his masterwork *On Liberty* (1859) an unmatched source of inspiration.

Yet the flipside of Mill's defence of individual liberty was his fear of 'the tyranny of the majority', particularly as it might be expressed and enforced through an expanded system of parliamentary democracy. Whilst accepting the principle of democratic reform, Mill essentially shared the prejudice of Plato, that a few wise men should in practice run the world. He wrote in 1835 that: 'The best government (need it be said?) must be the government of the wisest, and these must always be a few.

The people ought to be the masters, but they are masters who must employ servants more skilful than themselves.' Looking back later on the era of reform, Mill recalled how his wife Harriet Taylor (a champion of women's suffrage) and he had growing doubts about democracy because they 'dreaded the ignorance and especially the selfishness and brutality of the mass'.[30]

By 1859, as he was writing *On Liberty*, with further electoral reform on the way, Mill had come to accept another Platonic prejudice – that some voters should be treated more equally than others, not on the basis of their property, but because of what he called their 'proved superiority of education'. Under Mill's weighted voting system everybody, including women, would get one vote, but the educated classes would have more than one. This 'superiority of weight' would be 'justly due to opinions grounded on superiority of knowledge'.[31]

Even Tocqueville, the European champion of American democracy, feared that the industrial masses of brutal Victorian cities such as Manchester would threaten the democratic system. He suggested that governments would need to create an armed force which 'while subject to the wishes of the national majority, is independent of the peoples of the towns and capable of suppressing their excesses'.[32]

When the suffrage was eventually expanded, it often revealed popular sentiment to be so far from that of the high-minded elites that the authorities sometimes attempted to reverse the extension of democracy to the people in order to 'suppress their excesses'.

In the 1880s, for example, the expansion of the suffrage in the Austro-Hungarian empire facilitated the emergence of mass parties of a nationalist, 'pan-German' and generally anti-Semitic character, in an explosion of pent-up fervour. When

Karl Lueger, leader of the anti-Semitic Christian Social Party, was elected mayor of civilised Vienna in 1895, the Emperor Franz-Joseph refused to ratify his appointment for two years. The high-minded Viennese bourgeoisie rallied to the side of imperial diktat against the elected Lueger. As Carl Schorske, author of *Fin-de-Siècle Vienna: Politics and Culture*, says, 'Even the progressive Sigmund Freud, who in his youth had, like Beethoven, stubbornly refused to show respect for the emperor by doffing his hat, now celebrated Franz-Joseph's autocratic veto of Lueger and the majority's will.'

Meanwhile, the Reichsrat (national government) 'fell into such hopeless discord that the emperor had to dissolve it and establish government by decree ... The liberals could, however ruefully, only welcome the change. Their salvation lay henceforward in a retreat to Josephanism, an avoidance not only of democracy but even of representative parliamentary government, which seemed to lead to only two results: to general chaos or to the triumph of one or another of the anti-liberal forces.'[33]

Ultimately, Lueger was displaced by the new working-class movement of the Social Democratic Party, but not before the liberal elite had been confirmed in its fear 'not only of democracy but even of representative parliamentary government'.

Throughout the nineteenth-century era of expanding the franchise, then, the future of democracy remained in question. One pessimistic account even suggests that: 'If the South had won the American civil war, or if the United States had disintegrated politically during the 1850s and 1860s, the career of capitalist democracy in the West would probably have been a relatively short one.'[34] In the end, however, the capitalist democratic mainstream defeated both the slaveholders of America and the radicals in Europe, and emerged with an expanded but contained form of democracy that left the elites in control.

## The workers – and women – take centre stage

In the early twentieth century, the masses finally exploded onto the Western political stage. The more educated and organised working classes demanded not just better living standards but more of a say in society. In Britain, the formation of the Labour Party in 1905 soon gave the trade union movement its own voice in parliament for the first time, and led to increased demands to extend the right to vote to the working classes.

Meanwhile the militant suffragettes in Britain were campaigning to break the final taboo and win the vote for women. This was a key issue in the development of British democracy. The automatic denial of the vote to even educated and propertied women had been a basic rule of every formally democratic state since ancient Athens. Attitudes towards enfranchising all women would be an accurate measure of how far democracy had advanced.

There had been repeated attempts to legislate for votes for propertied women since 1867, when John Stuart Mill MP tried and failed to amend the Second Reform Act to include women in its provisions. In 1897, a Bill was actually passed by 71 votes; in 1904, by 114 votes; in 1908, by 179. But the government would not provide time in parliament to get such a law through. On one occasion Liberal prime minister William Gladstone threatened to ditch an entire reform bill unless his MPs backed down on their support for female suffrage. They did as they were told.

The vehement arguments raised against extending votes to women sound like exaggerated versions of the case against giving unpropertied men the vote. As one Tory MP summed it up in an 1870 debate, 'the male intellect is logical and judicial, the female instinctive and emotional'.[35] If working-class men were still considered too irrational, unthinking and emotional to be safely entrusted with the franchise, how much more dangerous would it be to treat women as fit to vote!

Led by the Pankhursts and their Women's Social and Political Union, the campaign for women's suffrage in the early twentieth century originally had close links with the growing labour movement. The Trades Union Congress had formally supported universal suffrage since its formation in 1868. However, before long the WSPU leadership sought to distance itself from the increasingly militant working-class movement, and to appeal to the governing parties to grant the vote to respectable middle-class women.

Even as the suffragettes themselves launched militant protests and 'terrorist' actions, enduring terrible punishments, they sought to contrast their campaign with that of striking workers. In March 1912 when Emmeline Pankhurst was arrested for throwing stones at No. 10 Downing Street, she made a long statement at the police station protesting that her actions were a 'fleabite' compared to the miners then on strike who were 'paralysing the life of the community' despite some of those men having the vote. 'If we had the vote,' she insisted, 'we would be constitutional.'[36] It was as if the suffragette leadership had internalised the argument against universal suffrage, and accepted that the vote was best suited to those property-owners of a respectable temperament. Critics said the WSPU slogan should be, not 'Votes for Women' but 'Votes for Ladies'. When the First World War broke out the leading suffragettes backed the war effort and suspended their campaign. They were rewarded in the 1918 Reform Act – with votes for propertied women over thirty years of age.

By contrast the radical wing of the Pankhurst family, Sylvia, continued to campaign for universal suffrage, for workers' rights, and against the First World War. As she said, 'the brave old reformers did not want the vote for merely academic reasons. They fought for it because they saw it as a way of giving all the people the power to free themselves from gaunt and urgent want, and to protect themselves from cruel exploitation and harsh injustice. They wanted to give every man an equal chance to share in controlling the destinies of the nation.

'Those old reformers,' said Sylvia Pankhurst, 'asked for no half-measure, suggested no paltering compromises, but demanded Universal Suffrage. They were determined to wring from the autocrats in power as much justice as they could, and not to abate their demands until they had got all they asked. Theirs is a spirit that we may well emulate.'[37] Hers is a spirit that all those struggling for democracy and social change have done well to emulate ever since.

## Unrest and revolution

By the early twentieth century the Western world's ruling elites might have been formally committed to democracy, but they could not trust the newly assertive masses and struggled to accommodate popular demands for greater choice and control over their lives.

Then, from 1914, the mass slaughter of the First World War wreaked terrible damage on the self-image of the supposedly free and democratic Western states. Popular dissatisfaction with alleged democracies that could compel millions of ordinary people, many of whom were still denied a vote, to march to their deaths in the service of empire was rising across the West.

As a new era of unrest and revolution opened, liberal democracy was on the defensive, widely seen as an illusory and hypocritical system. Western leaders first sought to celebrate the Russian revolution of February 1917 that replaced the Czar with parliamentary government – not least because it improved the image of the Allies of Britain, France and Russia as the pro-democracy side in the war against the autocrats of the Kaiser's Germany and Austro-Hungarian empire (even though Germany, unlike Britain or France, already had universal male suffrage). But in October 1917, when Lenin's Bolsheviks led a popular uprising of workers for whom it seemed even Western-style parliamentary government was not democratic enough, the Western authorities recoiled in horror and then sent invading armies to back the White Russians against the Bolsheviks in the civil war, in a despairing and failed attempt to restore the old order. The fear of revolution in the West was to shape the strategies of both reform and repression in the post-First World War era.

After the war, the elites sought to rebuild the reputation of liberal democracy with more piecemeal reforms. In Britain by 1918, it was still the case that only 60 per cent of male householders aged over twenty-one had the vote, and no women at all. Under the old system millions of soldiers returning from the front would have no right to vote in the forthcoming general election – the first one to be held in the UK since 1910. The Representation of the People Act of 1918 extended the franchise to almost all men aged over twenty-one, and to women aged over thirty who met minimum property qualifications. Women aged twenty-one did not get the right to vote until ten years later in 1928. At every stage, the right to participate in the democratic process was extended bit by bit and begrudgingly. And when the ungrateful masses asked for more, the authorities were willing to dispense with the democratic role play – as in 1919, when Secretary of State for War Winston

Churchill sent soldiers and tanks onto the streets of Glasgow to support riot police struggling to put down mass working-class unrest in the Battle of George Square.

Half-hearted reforms mixed with repression could not save the discredited parliamentary parties of Europe. After the war, mass radical movements of fascism on the Right and Communism on the Left emerged to challenge liberal democracy. The elites blamed the masses for being misled of course. Yet it was the lack of legitimacy and authority of the formally democratic states, and the contempt which the rulers showed to their peoples, which gave these mass movements their appeal, rather than simply the bigotry or ignorance of the masses.

Hitler's Nazi Party was able at first to gain millions of votes from those reacting against the political and economic bankruptcy of the old parties of the Weimar Republic. Even then, the Nazis did not ultimately win power through the ballot box, but via a deal to appoint Hitler Chancellor with the German establishment led by President Hindenburg, who did not trust the old parliamentary parties to contain the threat from the Communists and restore order to the new mass society.

The suggestion one hears these days, that the Nazis were able to seize power because of German democracy, turns history on its head; it was the anti-democratic instincts of Germany's aristocratic elites that led them to install Hitler in power in January 1933, at a time when the Nazis were losing support through the ballot box. As Hitler himself reportedly described later, in late January 1933 the German Chancellor and former general, Kurt Schleicher, had asked President Hindenburg 'for plenipotentiary powers to set up a military dictatorship' to restore order and crush the Communists. The President 'had refused and stated that he proposed inviting

Adolf Hitler, in the role of leader of a national front, to accept the Chancellorship and to form a government'.[38] The German establishment chose the Nazis rather than the military as their preferred weapon to usurp democracy and defeat the opposition.

The Western elites' fear and loathing of the masses in the inter-war years showed how little faith they had in the democratic system that they claimed to champion. American writer H. L. Mencken, an admirer of Nietzsche and technocracy, spoke for many when he famously declared his contempt for democracy and the people in the land of the free. 'Democracy,' wrote Mencken, 'is the theory that the common people know what they want, and deserve to get it good and hard.' In a 1919 column attacking tabloid newspapers aimed at 'near-illiterates,' Mencken spelled out the anti-democratic prejudice of his age: 'No one in this world, so far as I know – and I have searched the records for years, and employed agents to help me – has ever lost money by underestimating the intelligence of the great masses of the plain people. Nor has anyone ever lost public office thereby.'[39] For the elitist of the age democracy was a trick played by politicians to exploit the unintelligent 'great masses'.

Yet it is striking how similar attitudes penetrated not just the conservative establishment but the ranks of the liberal Left and cultural progressives.

Beatrice Webb, leading light of the Fabians and one of the intellectual founders of the Labour Left, wondered, 'If the bulk of the people were to remain poor and uneducated, was it desirable, was it even safe, to entrust them with the weapon of trade unionism and, through the ballot box, with making and controlling the government of Great Britain with its enormous wealth and far-flung dominions?'[40]

Intellectuals particularly despised the ignorant masses and the idea that they should decide the future of society through the ballot box. This attitude was articulated by H. G. Wells, author of *The War of the Worlds*, in his book *The Shape of Things to Come*, published in 1933, the year Hitler came to power. Looking back from an imagined future, Wells sees the twentieth century in 'the dustbin of history'. Democracy has been exposed as a 'shabby and threadbare religion' devoted to the worship of 'that poor invertebrate mass deity of theirs, the Voter'; a 'Divinity altogether too slow-witted for the urgent and political and economic riddles, with ruin and death at hand'.

For Wells, the only hope for the world was a 'technocracy' – rule by experts. But even that could only work once national barriers were removed and replaced by what Runciman describes as 'a technocratic world state, founded on efficiency and education, able to manage nature, neutralise religion, and channel human impulses in a healthy direction (the tide of porn that threatens to sweep the planet is finally reversed). The Age of Frustration is over.'[41] Sounds rather like the sort of technocratic, anti-democratic, dehumanising supranational state that some would like the EU to be today ...

## After the war: democracy by default

Yet after the Second World War ended, the West was surprised to find itself enjoying a revival of the standing of democracy. After the defeat of fascism and the revelations of the horrors of the Nazi Holocaust, nobody who wanted to be taken seriously could openly espouse anti-democratic politics. Even the new Stalinist state of East Germany gloried in the name of the German Democratic Republic, whilst former Nazi and fascist supporters on the German and Italian right converted to the new carefully named Christian Democratic Parties.

These cosmetically democratic measures, however, could not disguise the lack of confidence or belief in democracy among the ruling elites. This was obvious to all in the Soviet bloc of Eastern Europe.

Yet in the West too, the elites remained wary of indulging their apparently triumphant democratic system too far. In a 1947 speech to parliament, former wartime Tory prime minister Winston Churchill summed up the prevailing attitude: 'No one pretends that democracy is perfect or all-wise. Indeed it has been said that democracy is the worst form of government except for all those other forms that have been tried from time to time.' Churchill's lukewarm feelings for democracy contrast with the old political aristocrat's passionate defence of King, country and Empire. He let the mask slip further when suggesting that: 'The best argument against democracy is a five-minute conversation with the average voter'.[42]

The post-Second World War democracies of the West blamed the fickle masses for their own past failures. Political elites sought new ways to insulate themselves against a revival of popular pressure from the untrustworthy masses. They allotted new powers to expert commissions, constitutional courts and other unaccountable bodies.

This process was particularly pronounced in a new state such as West Germany, created by the occupying powers out of the rubble of the defeated Reich, where the authorities could invent a new constitution and political system from scratch to keep the German people at a safe distance from power. The result was not all that more democratic than its Stalinist neighbour to the East.

The top-down steps to create more Europe-wide institutions in the post-war era were similarly motivated by a desire to erect safety barriers between the *demos* and power. As the next chapter argues, the organisation that would become the European Union was from the first an attempt by Europe's political classes to find a powerful haven from their disaffected electorates back home.

Even in America, where democratic elections had continued throughout the war and the system appeared stable, the political class took steps to increase the centralised power of the state over society. Hence the special war powers granted to the US presidency in times of national emergency, and normally withdrawn at the end of the war, remained in place after the peace and through the Cold War era, an increase in undemocratic power that would become known as the Imperial Presidency. In 1964, for example, the Gulf of Tonkin resolution empowered Democratic President Lyndon Johnson to deploy US forces in Vietnam and South-east Asia as he saw fit – a transfer of the power of war from Congress to the White House of a sort that Abraham Lincoln had once suggested would turn a president into a king.

In the post-war era, fears about democracy were far from confined to traditional right-wing authoritarians. Now the liberal Left became increasingly wary of giving too much power to ordinary voters, supposedly misled by demagogues and the mass media and hooked on consumerist advertising. As the intellectual Left lost contact and support among the masses, it lost faith in democracy.

In America, the Left blamed the supposedly gullible masses for the McCarthyite anti-Communist witch-hunts. In the UK and Europe, a Left that was losing touch with the working class and turning on its popular base would help pioneer the displacement of politics into the undemocratic world of Euro courts and commissions.

Some forty-five years after the Second World War, Western capitalist democracy arguably reached its highest point of historical supremacy with the end of the Cold War and the collapse of the rival Soviet bloc. Yet even after the fall of the Berlin Wall, there was little evidence of any renewed faith in democracy among the rulers of the Western world. In 1989 American author Francis Fukuyama's 'End of History' thesis was hailed as a statement of the historic triumph of liberal democratic capitalism. Yet there was little real triumphalism in Fukuyama's argument. He based his case rather on the fact that all the alternatives had been discredited and collapsed. It was hardly a statement of deep commitment to or faith in the democratic cause. Western democracy was the winner by default. When Fukuyama expanded his thesis into a 1992 book, the full title became *The End of History and the Last Man*. He was at least half-right; the West had won by being the last man standing. We were soon to be reminded, however, that the history of the struggle for democracy never ends.[43]

Since then, what passes for Western democracy has increasingly been exposed as an empty shell. The D-word has been redefined so far that it means little to many of those it claims to represent. The spread of discontent, disengagement and disaffection across Western societies has revealed the yawning chasm between power and the people in whose name our rulers rule. The reaction against this discontent has brought much of the old elitist fear and loathing of the masses back into the public arena. *Demos*-phobia is back in fashion.

We have come a long way, not just from the *demokratia* of Athens but from the democratic aspirations of those who fought for freedom in the modern world. From direct democracy, to no democracy, to representative democracy and now unrepresentative democracy, one constant has been the efforts of those who see democracy as a dirty word to contain, redefine and hence to neuter it.

Those who demand democracy have been tolerated so long as they are prepared to stand quietly in the corner and take whatever they are given. But history suggests that democratic rights have been won in the West when people refuse to obey and find new ways to bring the fight for living democracy to the centre of the political stage.

The name of Karl Marx is not always thought of as synonymous with democracy, thanks to its association with Stalinist tyranny. Yet the young Marx was clear about the importance of the struggle for democracy, not just as a political system but to 'transform society into a community of men' with a 'higher purpose'. Back in 1843 Marx wrote of his fellow Germans that 'freedom, the feeling of man's dignity will have to be awakened again in these men. Only this feeling, which disappeared from the world with the Greeks and with Christianity disappeared into the blue mist of heaven, can again transform society into a community of men to achieve their highest purpose, the democratic state.'[44]

Today, awakening the history-making spirit of freedom in order to 'transform society into a community of men' and women must mean standing not only against old-fashioned tyrants, but also against those who would deny democratic freedoms in the degraded name of democracy itself. And none more so than the EU, as the next chapter examines.

# 4

# For Europe –
# against the EU

A basic point has often been lost sight of in the bitter debates about the future of Europe, the UK and Brexit. Europe and the European Union are not the same thing. At all. That is why, for more than a decade, some of us who believe in freedom and democracy have argued 'For Europe – Against the EU'.

Europe is an historic, dynamic continent of nations and peoples, of which we in Britain have been and always will be an integral part. What Europe 'means', what it should stand for, has been argued and fought over down the centuries. But if there is one fundamental value we might think of above all as European, it is surely freedom.

It is the freedom of political choice that was first glimpsed in ancient form among the citizens in the agora of Athens, first gained a modern voice some 2,000 years later in the European Renaissance and Enlightenment, and finally became the foundation of liberal democracy as we know it. This sense of freedom is one of the main reasons why many people in the UK like to think of themselves as 'pro-European', and why people from around the world want to come here.

By that standard, however, the EU itself is definitely not pro-European. Since its inception as the European Coal and

Steel Community in 1953, then the European Economic Community from 1956, to the European Union since 1993, one consistent value endorsed by the EU elite has been anti-democracy – the creation of a system that separates power and control in Europe from any expression of the popular will.

The EU's aim always was and remains not to 'represent' the peoples of Europe, but to constrain popular sovereignty and democracy. The European Union is not Europe. It is the anti-democratic union of Europe's political elites.

As the leading Spanish jurist Miguel Herrero de Minon wrote about the EU twenty years ago, 'the lack of "demos" ["the people"] is the main reason for the lack of democracy. And the democratic system without "demos" is just "cratos" – power.'[1] In the twenty years since then, the EU has gone further still in elevating the power of bureaucracy and technocracy over any sign of national sovereignty and popular democracy.

One feature of the debate around the UK's Euro-referendum in 2016 was the absence of any positive arguments for the EU. After sixty years of existence, it seemed that nobody had anything good to say about the 'European project'. All they could offer voters were scare stories and horror scenarios of how much worse things would be if Britain was outside the EU. The Remain campaign's slogan might as well have been 'You may not be EU-phoric about it – but Brexit means Brex-termination!'

The only semi-positive corollary of the negative case for Remain was the promise that the EU could be made to change for the better, especially if Britain fought for that change from the inside. In the words of Labour Party leader Jeremy Corbyn (a long-standing radical opponent of the EU who abandoned his principle and joined the conformist Remain camp when it mattered), we should vote and campaign for 'Remain and

Reform'. The idea is apparently to fight to make Brussels fill in its 'democratic deficit', to make the EU more democratic and accountable from within.

Yet as the left-wing British historian Eric Hobsbawm observed, reviewing the EU at the end of the twentieth century, it is 'misleading to speak of the "democratic deficit" of the European Union. The EU was explicitly constructed on a non-democratic (i.e. non-electoral) basis, and few would seriously argue that it would have got where it is otherwise.'[2]

Once we appreciate the inherently anti-democratic character of the European Union and its institutions, why bother trying to reform it? It becomes possible to see that as a lost cause before such a futile campaign even begins.

To suggest that we could reform the EU in a progressive, democratic way today is on a par with those who suggested it was possible to reform England's absolute monarchy, to make it less autocratic and meet the needs of the people, right until the morning of the execution of King Charles I in 1649; or those who proposed a reformed, more consensual form of British colonial rule at the moment when the Declaration of American independence was being signed in 1776.

As Tom Paine put it, arguing for ruling monarchies to be abolished rather than reformed and preserved in the revolutionary era of the eighteenth century: 'It will always happen when a thing is originally wrong that amendments do not make it right, and it often happens that they do as much mischief one way as good the other.'[3] There are moments in history where the only hope for freedom lies with beheading the tyrant, or kicking out the oppressors. Or leaving, and hopefully helping to break up, the democracy-eating monster that is the modern European Union.

## EU v 'illiberal democracy'

At first sight, the suave politicians of today's liberal–left EU lobby might appear to have little in common with old-fashioned reactionary anti-democrats. Yet despite the differences in style and language, they share some basic beliefs: that the experts and the better-educated always know what's best for the rest; that 'ordinary people' are not to be trusted with important decisions affecting society; that democratically elected governments are too prone to 'pander' to the unhealthy base appetites of the electorate; and that experiments in direct democracy such as referendums are the most dangerous of all, leaving the gullible masses prone to the appeals of populist demagogues.

Of course EU politicians and officials would never openly declare themselves to be the enemies of democracy. Instead, in the Orwellian Newspeak of the age, they present their techno-cratic rule as a defence of liberal democracy against the 'illiberal democracy' of some of Europe's elected governments.

The EU imagines itself to be the beacon of enlightened governance in a continent awash with xenophobic bigots being led astray by evil 'populists', from Austria and Hungary to Poland and the UK. Along with others in the political elite, EU officialdom has effectively redefined liberal democracy as a mindset, a set of manners, that simply means being decent and civilised rather than elected or accountable. Thus whatever the civilised suits of the EU say and do is democracy in action, and those who oppose them are a threat to democracy. According to this version of democracy, the decisions of civilised men and women in a Brussels committee room are deemed inherently superior to the verdicts of uncouth voters in actual democratic elections, which may well appear indecent by Brussels stand-ards and are best ignored or even overturned.

EU leaders enforce their idea of democracy by diktat. For example, at the start of 2012, twenty-five of the twenty-seven heads of state at a European Summit agreed to a new compact imposing binding limits on budget deficits, with the prospect of punishment by the European Court of Justice for those that broke the new rules. German chancellor Angela Merkel warned the weaker members of the Eurozone that: 'The debt brakes will be binding and valid forever. Never will you be able to change them through a parliamentary majority'.[4]

Just in case there was any doubt about the powerlessness of a mere parliamentary majority, shortly before that, in November 2011, the elected governments of both Greece and Italy had effectively been dumped out of office and replaced by unelected technocrats. These new governments were endorsed, not by Greek or Italian voters, but by a powerful Troika of the European Commission, the European Central Bank and the International Monetary Fund, which had assumed control of their national economies to force through punitive measures in response to the financial crisis.

This sort of thing, one might think, exposes the EU as a dictatorial power. Yet the Euro-elites see no apparent contradiction between deposing elected governments and posing as the guardians of democracy.

At the same time as the Troika was taking over Greece and Italy, the European Commission sent severe warnings to Hungary's elected right-wing prime minister Viktor Oban, charging him with introducing 'undemocratic' laws – that is, laws which were contrary to the tastes of the Commission. Oban's new constitution had dared to put Hungary's central bank under the control of the government rather than technocrats, and to suggest that constitutional judges might be accountable to parliament rather than just their own consciences and egos. As James Heartfield, author of *The European Union and the End of Politics*, noted, to follow such events is to 'Step through the looking glass into the EU-world where the rule of the people is dictatorial, but the rule of unelected experts is democracy.'[5]

The EU elite has since become more open in its war on 'populism' and the 'illiberal democracy' it supposedly breeds. Faced with the possibility of the Freedom Party winning the Austrian presidential election in 2016, EC President Jean-Claude Juncker declared that such a result would not count in Brussels. 'There is no debate or dialogue with the far-right', he warned.[6] Juncker was telling the Austrian electorate that if they exercised their democratic right to vote for the 'wrong' parties, the EC would go into a huff and refuse to talk to their chosen head of state.

Anybody with a democratic bone in their body might understand that, if you want to challenge the politics of the far right in Austria or elsewhere, then debate is precisely what is required. More free speech, rather than less, is the potential solution to changing the minds of Norbert Hofer's supporters. For Juncker and the EU oligarchy, however, public debate has no part in their idea of democracy.

Meanwhile there are influential cheerleaders who urge the Euro-authorities to go further still in overthrowing the will of the people in 'illiberal democracies' in the name of liberal democracy. Political Science Professor Jan-Werner Müller of Princeton University wants to see the EU intervening more forcefully in member-states with right-wing populist governments such as Hungary and Poland, to pursue a policy of 'supra-national militant democracy'. Professor Müller does not demand that Brussels wage its 'militant' crusade against elected EU governments with air-strikes (not yet, anyway), but through a powerful 'new democracy watchdog' of Eurocrats who could intervene to bring the upstarts to heel.[7]

Those demanding that the EU wage war against 'illiberal democracy' to defeat the Right miss the point. It is precisely the anti-democratic practices of the EU which gave rise to the right-wing populist opposition in the first place. The fact that some far-right parties have made an electoral impact in Europe is not really an argument against democracy; it's an argument for more democratic debate and less of the diktat of the EU technocrats.

It has long been the habit of the Euro-authorities to suggest that any challenge to the EU is comparable to the fascists and right-wing extremists who attacked democracy in the twentieth century and plunged Europe into world war. As such, 'Euroscepticism' has been deemed automatically illegitimate and beyond the pale of respectable political debate. The response of the EU elites has been not to engage or argue with their critics, but to dismiss them and seek to silence them as a threat to Europe's peace and liberal democracy. This has back-fired in quite spectacular fashion.

A decade ago, social anthropologist Maryon McDonald conducted in-depth interviews with EU officials in Brussels. She was struck by their intolerance of anything more than mild criticism. Any serious questioning of the EU would be branded, by definition, as dangerous right-wing extremism. 'Since the 1970s especially,' wrote McDonald, 'it has become increasingly difficult in Europe to criticise the EU without appearing to be some lunatic right-wing fascist, racist or nationalist, the one often eliding with the other, or simply the parochial idiot of Little Britain.' The consequence, she observed, had been the opposite of what the Eurocrats imagined: 'The serious side of this is that the EU has, quite literally, encouraged neo-nationalist racism in Europe. That has often seemed the only space available in which criticism of the EU has been possible. A new space of serious criticism is therefore badly needed.'[8]

By automatically condemning its critics as racist, the EU has managed to enhance the reputation and authority of often crankish right-wing movements more than they could ever have managed alone. Often the only outlet for popular discontent with the EU has been through these right-wing parties – especially since the European Left abandoned its own commitment to democracy and began acting like EU groupies. If you seek an explanation for the rise of 'populism' in Europe, look not at some imagined innate racism of the people, but at the ingrained anti-democracy of the European Union and its refusal to countenance criticism of its rule or serious debate on the future of Europe.

## 'Deep distrust of popular sovereignty'

Apologists for the EU will often claim that its formation after the Second World War was an attempt to save Europe from further terrible conflicts. What is less often discussed is that European elites effectively blamed mass politics and the volatility of popular opinion for the conflagrations of the twentieth century. They saw national sovereignty as the primary cause of war – because of the 'hegemony' of nationalist politics over the peoples of Europe. In this one-eyed worldview, mass politics had led to war. To those seeking to build a new peace in Europe from the top down, then, popular democracy was part of the old problem, not the new solution.

The founders of the EU sought to create a new system that would allow them to manage Europe's affairs whilst being insulated from the pressure of the masses. They would defend the overarching idea of democracy in Europe – by protecting it against the inconvenience of actual democratic politics in nation states. Thus was the EU's uneasy relationship with democracy formed from the start.

The champion of 'supra-national militant democracy' cited above, Professor Müller, admits that: 'Insulation from popular pressures and, more broadly, a deep distrust of popular sovereignty, underlay not just the beginnings of European integration, but the political reconstruction of Western Europe after 1945 in general.' Driven by their mistrust of the masses, the new European elites 'fashioned a highly constrained form of democracy, deeply imprinted with a distrust of popular sovereignty – in fact, even a distrust of traditional parliamentary sovereignty'.[9]

Constraining democracy in post-war Western Europe meant creating systems that had formal elections and elected representatives, but at the same time would take democratic politics and the populace out of the business of government. In nation states that were rebuilt or invented after the devastation caused by the war, such as West Germany, new constitutions shifted powers away from parliaments into the hands of judges, central bankers and officials.

The West German constitution, imposed by the occupying Allied powers but with crucial local support, set up a new Constitutional Court sitting above elected institutions with the power to rule out laws passed by parliament. In the words of one expert, 'Germany was explicitly created as a *wehrhafte Demokratie* (defensive democracy), one with structures designed to "protect Germans from themselves"'.[10] With popular democracy widely blamed for the rise of Hitler, the German elites (who had actually put the Nazis in power) were happy to entrust more power to sober judges rather than the impassioned people.

Key to the wider process of 'depoliticising' government in Western Europe in order to 'protect Europeans from themselves' was the formation and forward march of the bureaucratic institutions of what would become the European Union. Debate and decisions were to be shifted out of the arenas of national democracy and into the committee rooms and courts of the EU.

The signals were clear back in 1951, when the leaders of the six founding nations – West Germany, France, Italy, the Netherlands, Belgium and Luxembourg – signed the famous Europe Declaration which set in train the creation of the European Community and then the EU. It stated that the signatories 'give proof of their determination to create the first supra-national institution and that thus they are laying the true foundation of an organised Europe'.[11] The prefix 'supra', from the Latin, means above, over or beyond. The 'supra-national' institutions of the embryonic EU would operate over and above national politics, and beyond the reach of the citizens of any nation state. The aim was to create a new 'organised Europe' managed not in the public realm of democratic politics, but in the closed world of top level Euro-bureaucracy and diplomacy.

From the 1950s until today, the clear intention of the European political elites has been to create a supranational form of unity above and beyond the reach of national parliaments. This was never, however, a case of an alien Brussels empire somehow conquering the major nations of Europe with only an army of paper-pushers. National political elites willingly signed up to the process of political unification, in order to give themselves more protection from political scrutiny and democratic accountability at home. As the Brussels correspondent Bruno Waterfield writes, 'The EU has evolved, not as a federal super-state that crushes nations underfoot, but as an expanding set of structures and practices that have allowed Europe's political elites to conduct increasing areas of policy without reference to the public.'[12]

From the establishment of the European Economic Community (EEC) or 'Common market' in 1958 through to the launching of the European Union under the Maastricht Treaty in 1993, the political elites of major European nations sought to create an integrated, technocratic system of government as a way of removing big issues from their own democratic national politics. Handing authority to the EU became a way of avoiding being held accountable at home.

Even in the UK, which has often enjoyed an uneasy public relationship with the EU, the national government has often been keen to use Euro-bureaucracy in order to bypass political debate at home and implement troublesome or unpopular measures. In the early 2000s, for example, the New Labour government of Tony Blair suffered setbacks in the UK parliament over its plans to give state bodies even more surveillance and snooping powers. In response, Labour home secretaries David Blunkett and Charles Clarke went to Brussels and Strasbourg and urged the EU to get on with enforcing Europe-wide measures enabling security forces to access and share telecommunications data.

Nor is it only Labour governments who have used the EU to dodge political problems in the UK. In the mid-1990s, the supposedly Eurosceptic Tory government turned to Europe to avoid embarrassment over the tricky issue of allowing gays to join the British Army. A leaked memo revealed that the Conservative government was using different tactics on two fronts. In public, the Tories would remain staunchly in favour of the ban on openly gay soldiers, to assuage their traditional support. In private, however, the ministry of defence knew that change must come. It was quite content that the European court would sooner or later overturn the ban and save the Tories the embarrassment of tackling the tricky issue.[13]

Even under prime minister Margaret Thatcher, often depicted as the modern-day Boudicca defending Britain against the Brussels Empire, the Conservatives were sometimes keen to lean on European institutions to help them avoid domestic political pressures. Thatcher's chancellor Nigel Lawson saw signing the UK pound up to the European Exchange Rate Mechanism (ERM) in the 1980s as a way of maintaining a grip on finances whilst avoiding responsibility for the unpopular results; submitting to an 'externally imposed exchange rate mechanism', he thought, would help avoid 'the political pressure for relaxation … as the election approaches'.[14] In other words, if money's too tight to mention blame the German Bundesbank, not the British government!

## EU citizens, like it or not

At least until the 1980s, trying to win public approval for their project was not high on the Euro-elites' priority list. Their attitude was summed up by Pascal Lamy, top aide to then-European Commission president Jacques Delors: 'The people weren't ready to agree to integration, so you had to get on without telling them much about what was happening.'[15] Instead they operated a system of 'implicit consent'; you could simply assume that the people approved of whatever you were doing, unless they explicitly said they didn't. Which was unlikely, since you never actually asked them to express a preference.

The EU-builders more or less got away with this condescending attitude to the people during the early decades of post-war economic prosperity. They were also aided by the spirit of Western political unity enforced during the Cold War. But by the late Eighties, as they moved towards launching the more integrated European Union, they were faced with conditions of economic crisis, the end of the Cold War and growing political instability. Now the Euro-elites felt the need to seek some form of greater public legitimacy for their grand 'European project'.

How to win public support for their technocratic project without actually extending democratic power? There had been direct elections to the European Parliament since 1979, but it already seemed clear that there was little public enthusiasm for that toothless talking shop, with fewer voters bothering to turn out for the elections.

So was born the artificial project of creating 'European citizens', designed to make us all feel part of the EU without granting us a stake in its decision-making process. This is a very peculiar sort of citizenship. To be a citizen normally means to be an equal member of a nation state. Citizenship is a status that has had to be won in the past by democratic political movements, often forcibly wrested from ancient regimes; think of the cry '*Aux armes, citoyens!*' from the French revolutionary anthem, 'La Marseillaise'.

By contrast, EU citizenship was invented and imposed entirely from the top downwards. The Maastricht Treaty which formed the European Union in 1993 simply pasted 'Citizenship of the Union' into the original Treaty of Rome that founded the European Community. It declared that everybody from a member-state was now an EU citizen, who 'shall enjoy the rights conferred by this Treaty and shall be subject to the duties imposed thereby'. The 'rights' were the 'Four Fundamental Freedoms' – free movement of goods, services, people and capital within the EU. But as Cris Shore, author of *Building Europe*, observes, 'What exactly the duties are, Article 8 does not say, nor is there any indication in what direction the provisions may be leading – a point that has prompted criticisms that the governments of Europe have effectively signed a blank cheque for their citizens.'[16]

Crucially, as Shore concludes, 'Whatever rights may have been conferred by Union citizenship, the right to choose whether to claim it was not an option for those qualifying for it.' This pseudo-citizenship is obligatory. So long as yours is an EU member-state, then you are an EU citizen, like it or not, enjoying the privilege of those unspecified 'duties' but with no real democratic rights.

EU citizenship was invented by the continent's political elites in order to address their crisis of legitimacy and try to forge a sense of 'Europeanness'. But it has nothing to do with being a 'citizen of Europe' with an equal stake in society. Enjoying EU citizenship is more like being a subject of a king, who assumes the right to rule over us whether we like it or not.

Without the democratic accountability that is the basis of meaningful citizenship in modern nation states, the EU authorities have had to try to compensate through creating cultural symbols of a European identity.

The EU flag, with its twelve stars, was adopted and first flown in 1986. EU history books have been rewritten to emphasise the importance of post-war European unity. An official body called the Eurostat office is now busy compiling cross-EU polling figures that are supposed to represent something called 'European public opinion', a previously unheard-of phenomenon.

Yet ask for evidence of any public sense of EU citizenship, and the best its apologists can normally do is talk about the popularity of the Eurovision Song Contest, student Interrailing, or support for 'Team Europe' in the Ryder Cup golf contest with the US. These rather pathetic bits of cultural ephemera are a poor substitute for having a say through democratic citizenship.

Even when it tries to make a connection with us 'EU citizens', the Euro-bureaucracy cannot help treating its subjects with disdain. Back in 1993, at the launch of the European Union, a major report by the Belgian MEP and former Euro commissioner Willy De Clercq set the tone. De Clercq admitted that European integration was 'a concept based far more on the will of statesmen than on the will of the people'. The answer was to stop trying to explain the Maastricht Treaty; that was 'far too technical and remote from daily life for people to understand', and should be left to the experts to decide. Instead, they must promote the EU brand to its simple people as a 'good product' with 'a human face – sympathetic, warm and caring'.

European institutions, said De Clercq, must be 'brought closer to the people, implicitly evoking the maternal, nurturing care of "Europa" for all her children'.[17] The idea of the stony-faced EU as a caring parent who knows what's best for all of us Euro-kiddies points to the infantilising effect of undemocratic rule.

That the EU parent still assumes the 'implicit consent' of Europe's peoples, even when they explicitly withdraw it, was clear in the process of launching the euro and European Monetary Union. In April 1998, pollsters found that no fewer than 62 per cent of the German public were opposed to the abolition of their national currency, the Deutschmark, and adoption of the Euro. At the same moment, considerably more than 62 per cent of the German parliament voted for EMU – endorsed by 575 votes to just 35. It was a powerful illustration of what Jacques Delors called the 'benign despotism' of the

European elites, to which national parliaments have generally bent the knee.

By May 2013 José Manuel Durão Barroso, then president of the European Commission, was admitting that 'implicit consent' for advancing integration was no longer good enough: 'Europe has to be ever-more democratic. Europe's democratic legitimacy and accountability must keep pace with its increased role and power.' In practice the grand talk only resulted in another PR campaign to improve the image of the EU 'brand'.

The reality of the EU's attitude to 'democratic legitimacy and accountability' was spelled out in more honest fashion in response to the UK vote to leave. Speaking in Berlin a few months after the Brexit referendum result, as talks about the terms of future relations between the UK and the EU grew nearer, Europe's most powerful leader, German chancellor Angela Merkel, made clear the need to take a firm line to protect the future of the Union: 'If we don't insist that full access to the single market is tied to complete acceptance of the four basic freedoms, then a process will spread across Europe whereby everyone does and is allowed what they want.'[18] Everybody being allowed to do what they want! That is anathema to the EU spirit of conformism and control.

## Member-states, not nation states

Nothing brings out the EU elites' fear of popular politics more than mention of a referendum. This long pre-dates the Brexit debate. For years, they have done all they can to avoid holding referendums on European issues – and to ignore or overturn referendum results that are not to their liking.

Remember what happened when they tried to impose a grand new centralised EU Constitution on all members? In 2005, the constitution was rejected in referendums by first

French and then Dutch voters. To avoid further humiliation, Europe's leaders withdrew the plan. Or rather they re-labelled it as the Lisbon Treaty, which could just be agreed by EU leaders among themselves at a top-table summit. It need only be put to a referendum in Ireland, where the constitution explicitly required it.

French president Nicolas Sarkozy explained the explicitly anti-democratic thinking behind this paper-shuffling: 'France was just ahead of all the other countries in voting No. It would happen in all Member States if they have a referendum. There is a cleavage between people and governments ... A referendum now would bring Europe into danger. There will be no Treaty if we had a referendum in France, which would again be followed by a referendum in the UK ...'[19] Even when they recognise that there is 'a cleavage between people and governments', the response is not to engage the people in a debate and a vote, but to take a cleaver to democracy and ignore their wishes.

When that sole referendum on the Lisbon Treaty took place in Ireland in June 2008, the Irish people showed what they thought of such EU chicanery by voting to reject the Treaty by 53.4 per cent to 46.6 per cent. The response of the EU elites, normally allergic to any mention of a referendum, was to demand that Ireland hold another vote to produce the correct answer. After more than a year of moral blackmail and bullying, that second referendum was held in October 2009, when Irish voters accepted the Treaty by 67.1 per cent to 32.9 per cent. A triumph of liberal democracy, EU-style.

The EU is said to be 'an ever closer union among the peoples of Europe'. In reality it operates as a private union of Europe's political elites, from which 'the peoples' are excluded. Look at the main institutions where the EU does its business, in an atmosphere of secrecy and public silence where the rooms are as free of the air of democracy as they are of tobacco smoke.

The core of the EU's business is done through COREPER – the Committee of Permanent Representatives – a gathering of senior national officials which handles 90 per cent of EU legislation. Its proceedings are treated as state secrets, its documents usually classed as 'non-papers', which means they cannot be accessed by the press or the public despite the EU's supposed open information rules.

COREPER does the spadework in preparation for the regular meetings of the Council of the EU, which brings governments together to agree on Euro-legislation, usually behind closed doors. Much of this has already been decided in the Council's hundreds of committees and working groups, all of which operate in secret.

The big showcase for the EU is the regular meeting of the European Council, usually described as a 'summit' of European leaders. There is no public record of what is said in there, just a set-piece media photo-opportunity and a summit communiqué prepared by COREPER. This document, known as Council Conclusions, binds governments to what has been agreed, regardless of what happens in their domestic politics or elections between meetings. As Bruno Waterfield says in a survey of how EU secrecy bypasses democracy, 'The Council Conclusions are a compact between leaders that overrides the relationship between voters and their governments'.[20]

This is a key point. Some might object, after all, that the European Council is a symbol of representative democracy, since it brings together the various elected heads of government and state. But it does not simply bring them together as independent sovereign governments. In the process, it transforms them into something else. As Chris Bickerton, author of *The*

*European Union: A Citizen's Guide*, argues, they cease to be the representatives of nation states accountable to their electorates and transmogrify into a new political entity: 'member-states' of the EU. These member-states draw their authority from their membership of the Union and seat at its top table. They become members of an exclusive club to which the public need not apply; their name's not down and they're not getting in.[21]

Then there is the European Commission (EC), the most commonly heard-from public face of the EU and the only body that can propose legislation. The EC is an unelected executive, which believes it is practising what one of its former presidents called 'benign despotism'. This bureaucratic body proposes and polices thousands of EU rules and regulations, in consultation with an army of expert officials who would not know a voter if they bumped into one at lunch at a Brussels restaurant.

For the historian Perry Anderson, 'What the trinity of Council, Coreper and Commission figures is not just an absence of democracy – it is certainly also that – but an atten- uation of politics of any kind, as ordinarily understood. The effect of this axis is to short-circuit – above all at the critical Coreper level – national legislatures that are continually confronted with a mass of decisions over which they lack any oversight.'[22] Rather than political issues to be debated and decided in national parliaments, major questions that affect domestic politics become treated as technical matters to be sorted and filed away in committees and secret diplomatic summits.

But what about the European Parliament? What's not demo- cratic about that? All those hundreds of MEPs are directly elected by voters in their nations, after all. They in turn then vote for the President of the Commission, and even have the power to remove the Commissioners if they collectively choose.

Well, it's a parliament, Jacques, but not as we know it. The European Parliament is not a legislature – it has no power to propose and pass laws. It does not elect a government, like the parliaments of European nations. It does not even have the power to choose where it sits, shuffling back and forth between Brussels and Strasbourg at the whim of the Commission. It offers rather drab, expensive and unconvincing democratic window dressing for a system where the real power emanates via bureaucratic diktat and secret diplomatic deals.

Any notion of the Euro Parliament as the 'democratic conscience' of the Union was surely dispelled during the corruption scandal that rocked the normally immovable EC at the end of the 1990s. Faced with the chance to sweep all of the discredited Commissioners away, the MEPs ducked the challenge. Cris Shore justifiably doubts 'whether there can ever be effective accountability of the Commission to the European Parliament, particularly given parliament's lack of accountability to its electors'. They are all members of the same *classe politique*. As one official commented at the time of the fraud scandal, 'MEPs can't really criticise the Commission because parliament shares the same institutional culture as the Commission. If there is less fraud in the Parliament it is because the European Parliament's budget is smaller'.[23]

## What's left?

Perhaps it should be clearer now why the pro-EU lobby finds it so hard to offer any positive or passionate case for 'the European project'. How could anybody feel passionately committed to such a technocratic system of governance? It would be like getting overexcited about your choice of accountant.

Passions and popular support tend to come with a belief in political ideas and ideals. By contrast, the EU elite lacks political ideals and does everything it can to depoliticise questions which might excite the interest of the public. It has no grand vision to project other than the passion-killing idea of a technocratically managed Europe of the committee men.

The petty acts of regulatory interference in everyday life for which the EU is notorious assume such importance because it stands for nothing more substantial. This ever-expanding catalogue of Euro rules and regulations, touching everything from the correct size of a fish to the permitted wattage of a hair dryer or vacuum cleaner, is partly the EU's attempt to justify its existence as a cross between a head teacher and a health and safety officer, instructing the Continent's childlike peoples for their own good.

The EU sucks the life out of politics and deadens democracy. All the more remarkable, then, that defending the EU has seemingly become such a principle for much of the Left in the UK and elsewhere in Europe.

It has become common to try to understand the debate on Europe along traditional Left–Right lines. The liberal Left is supposedly for the EU as a bastion of its cosmopolitan values, while the conservative Right is said to favour more nationally-based politics. From this point of view, those supporting Leave in the UK referendum were often automatically assumed to be Tories 'or worse'.

This simplistic version of events bears little relation to reality. In fact millions of Leave voters were working-class Labour supporters, or at least former Labour voters. One group who did almost wholeheartedly support Remain, however, were Britain's left-wing intellectuals and activists.

This represents a remarkable turnaround. There is a long left-wing tradition of opposing the EU and top-down European integration. In the 1975 referendum on whether the UK should remain a member of what was then the 'Common Market', two of the most prominent supporters of the 'No' camp were the leading left-wing Labour MPs Tony Benn and Michael Foot, who later became party leader. These leaders of the Left were so committed to getting Britain out of the 'capitalist club' in Europe that they were unfazed by finding themselves on the same 'No' side as the anti-immigrant right-winger Enoch 'rivers of blood' Powell.

Yet in 2016 left-wing Labour leader Jeremy Corbyn – a former acolyte of the late Tony Benn and long-term opponent of the EU – signed up to the Remain campaign, along with most of his party's MPs, members and celebrity fans. What happened in the forty years between to explain this change?

The shorthand version of what happened is that the Left in the UK and across Europe lost the political war at home, and so sought refuge in the EU. The defeat of the powerful trade union movement in the 1980s, most notably in the 1984–5 miners' strike, was followed by the fall of the Berlin Wall and the collapse of the Soviet Union. That not only destroyed Stalinism, but also dealt a heavy blow to all those on the Left whose politics rested on a paler version of Soviet-style state socialism.

Finding it harder to win an argument or connect with a working-class constituency in the UK, the Left seized upon the peaceful pastures of Europe's courts and commissions as a more fruitful field to work in. A turning point came when Jacques Delors, then president of the European Commission, spoke at the British Trades Union Congress in 1988. Margaret Thatcher's union-bashing Conservative government had won a third consecutive victory and the Left was in disarray. Delors offered them a vision of a more Left-friendly EU, embodied in the 'social chapter' of the forthcoming Maastricht Treaty that dealt with such matters as working hours and conditions. In response the TUC delegates gave the top Eurocrat a standing ovation, and a rousing chorus of 'Frère Jacques' – a slight contrast to the tabloid *Sun*'s 'Up Yours, Delors!' headline from the same time.

The British Left's 'turn to Europe' had nothing to do with any principled support for the EU. It was simply an opportunist attempt to escape their isolation and loss of working-class support at home by seeking well-heeled allies to work with in Brussels, Strasbourg and The Hague, lobbying EU officials and judges to impose 'progressive' measures that they could not win support for in the UK.

In this respect, it was the mirror image of what the despised Tories had done a few years earlier. At the time of the 1975 referendum, Iain Duncan Smith (later to become a short-lived Conservative leader and then Brexit campaigner) explained that his fellow Tories had joined the 'Yes' side because 'they were all convinced that the argument against socialism had been lost in Britain, and that they could only hope to stop its advance in Europe'.[24]

Leading Tories in the Seventies shelved their Euroscepticism and looked to Europe to reverse the Left's gains by encouraging market economics. From the 1980s the Labour Left took a leaf out of that Tory book on Europe and did much the same in reverse, when they felt that winning the argument and public support in 'Thatcher's Britain' had become a lost cause. Both sides were driven into the arms of the anti-democratic EU by their loss of support at home and loss of faith in democratic politics.

Now that supporting the EU has apparently become an item of left-wing faith, even turning somebody like Corbyn into a 'Remain and Reform' campaigner, it is worth recalling the rather different attitude struck by the late Tony Benn, the leading light on the Labour Left for many years before his death in 2014.

Some of that earlier generation of left Eurosceptics, such as Foot, saw the European Common Market in quite narrow economic terms as basically a bosses' club to support big business across the Continent. To his credit, Benn always viewed Europe from a wider political perspective. For him the primary issue was democracy – and its absence at the heart of the European project.

Benn recalled travelling to Brussels to meet Euro bigwigs as a minister in the Labour government in 1974, and feeling 'like a slave going to Rome' under the thumb of the emperor. As he explained to the Oxford Union in his later years, after the horrors of the Second World War Benn had shared the ambition of many to see Europe united in peace and harmony. However:

> When I saw how the European Union was developing, it was very obvious that what they had in mind was not democratic. I mean, in Britain you vote for the government and therefore the government has to listen to you, and if you don't like it you can change it. But in Europe all the key positions are appointed, not elected – the Commission, for example. All appointed, not one of them elected.
>
> And my view about the European Union has always been not that I am hostile to foreigners, but that I am in favour of democracy. And I think out of this story we have to find an answer, because I certainly don't want to live in hostility to the European Union but I think they are building an empire there and they want us to be a part of that empire, and I don't want that.[25]

Benn's response to the TUC's embrace of Delors and the European Commission as their potential saviours from the dark days of Thatcherism is even more to the point. In contrast to his comrades on the Left who were inviting the EC to impose 'progressive' rules and working regulations on the UK, Benn stood up for the principle of democracy, even if it meant sometimes getting stuck with the wrong government. Speaking in the House of Commons in 1991, as the debate on a federal Europe heated up with the signing of the Maastricht Treaty imminent, Benn distanced himself from the cross-party support for the new EU: 'Some people genuinely believe that we shall never get social justice from the British Government, but we shall get it from Jacques Delors; they believe that a good king is better than a bad Parliament. I have never taken that view.'[26] And neither should anybody on the Left or Right who pays more than lip service to the principle of democratic politics.

That other veteran left-wing opponent of the EU, the late Michael Foot, used to recall some advice he got from his father. Whenever you meet a man, Foot senior suggested, ask yourself: which side would he have been on at Marston Moor? That 1644 clash between the forces of the King and Parliament was a decisive battle in the English revolution. We might say that, regardless of how anybody voted in the EU referendum, which side you take now in the battle over the future of democracy is a defining political issue of our time.

## Open borders, closed minds

One issue that somehow still sustains the nonsensical notion of the EU as a progressive, open-minded body is that of migration. The notion that being pro-EU meant being for open borders, while voting Leave meant hating refugees, was one of the main myths peddled by liberal Remainers during and after the UK referendum.

This idea was underlined shortly after the Brexit vote, when EC president Jean-Claude Juncker declared that 'borders are the worst invention ever made by politicians' and called on Europe to support migrants. Such a liberal and progressive-sounding statement, however, is not what it might seem.

If the EU were really against 'borders', why would it have spent billions of euros building and protecting 'Fortress Europe' to keep out all those millions of migrants from the developing world? If the EU was serious about supporting migrants, might it not throw more than the occasional lifejacket to the thousands of them drowning in the Mediterranean Sea as they try to evade its stringent border controls?

No, the borders Juncker and Co despise are only national borders within Europe, and the popular politics which go on within them, as the Commission president made clear in that same speech. 'We have to fight against nationalism', said Juncker, '[and] block the avenue of populists.'

That's why EU immigration policy is all about depriving European nations of their right to determine their own approach to immigration. It involves imposing quotas on member-states and denying national governments the authority to control the flow of migrants into their society. It is about controlling, hobbling the nation state. Those like the Hungarian government who object to this Euro-imposition can then be condemned as racists and neo-fascists.

Once upon a time, the aim of liberal supporters of more open borders was to defend the human aspiration for freedom of movement and social mobility. By contrast in EU circles today borders are seen as problems, not because they control the movement of people but because they are under the control of a nation state. Juncker and Co don't celebrate migration because of any liberal conception of open borders being good for people. They do so because they believe national sovereignty is a bad thing that allows a people too much say in what happens in their nation. It is disdain for popular sovereignty and democracy, rather than support for migrants, that drives them to call for open borders within the EU (whilst fighting to defend its external borders to the last drowned non-European migrant). As Professor Frank Furedi writes, 'Juncker's animosity towards borders is inspired less by a love of migrants than by a loathing of the nation state.'[27]

Thus the EU technocrats have managed the considerable achievement of twisting an apparently anti-racist argument about borders into another illiberal restriction on democracy.

Those who want to be on the side of migrants need to have an open argument about the politics of immigration rather than hiding behind the EU. Whether you favour open borders or not, the prerequisite is that our society has democratic control of those borders to make our own choices, rather than having it controlled by the quota-policing bureaucrats of the EU. As ever, the best hope for open-minded politics is open, no-holds-barred political debate about controversial issues – the one thing to which the EU mindset is permanently closed.

## Bring down the wall

How does anybody who has glanced at its history still retain the illusion that it might somehow be possible to reform the EU in a democratic direction?

The common pattern through those decades is that, every time the 'European project' runs into trouble, the instinctive response of the EU elites is to push for even more top-down, undemocratic integration. In other words, every time there is a problem, their solution is to make matters worse.

The severe economic crisis in the Euro-zone, for example, and the terrible consequences it has inflicted on the people of a member-state such as Greece, might reasonably raise questions about the viability of European monetary union. For the leading politicians, bankers and technocrats of the EU, however, it proves only the opposite: that they really need an even more centralised financial system which can cut out the elected middle men altogether and give the mighty European central bank direct control of the Euro-minnows' economies.

It is said that a shark has to keep moving forward in pursuit of prey or it will perish. In similar vein, it seems that the EU has to keep advancing towards further elite control, devouring any morsel of democracy in its path, or lose its reason to live.

The EU is not Europe. European unity is a positive ideal. But the European Union is something else entirely. An idea of EU citizenship imposed on us all from the top down, without consent ever being asked or given, is very different from a unity forged by the peoples of the Continent in democratic debate. That unity must begin by recognising the sovereignty of nations and their right to decide.

The political divide on Europe today does not follow the old Left–Right lines. It is more a divide about democracy: do we stand for democratic decision-making through a defence of popular sovereignty, or will we accept the technocratic governance of the EU?

In terms of its own values of freedom, Europe would be far better off without the EU. There are parallels here with the Berlin Wall. When the Wall fell in 1989, there were widespread celebrations. Yet many on the Left were fearful of what would follow, warning that this could only benefit the political Right.

Some of us insisted, however, that whatever the short-term turmoil and trouble it unleashed, the Wall had to go if there was to be any hope of European progress and unity.

In that sense, the EU now looks rather like a Berlin Wall for the twenty-first century. The anti-EU movement may well be dominated today by those on the Right (that, as suggested above, is largely the pro-EU Left's own doing). Upheavals such as Brexit will bring uncertainty and instability, at least in the short term. None of that alters the fact that bringing down the EU is as historically necessary as getting rid of the Berlin Wall, if living democracy is to have a future in the freedom-loving Continent of Europe we share.

In the end, the negative argument that we might be worse off without the guiding hand of the technocrats is not a good enough reason to sacrifice democracy and bend the knee to the EU elites.

In November 2010, as an EU–IMF austerity package of spending cuts and tax rises was being imposed on Ireland, the Irish commentator Fintan O'Toole spoke to a huge crowd protesting in the snow outside Dublin's General Post Office. The question was not whether people would make financial sacrifices, he said – they made them all the time for their families. The question was who had the power to decide the destiny of the Irish people. 'We are here today to say that we are not economic units whose only function is to behave ourselves … We are not children who must take our medicine or be sent to bed without our supper … We are citizens. And we want our republic back.'[28]

Those of us in the UK, who might never have had a republic in the first place, would do well to endorse the democratic spirit of his words.

# 5

# Some popular arguments against popular democracy

**1 'If you let people decide, they'll vote to bring back hanging and kick out migrants'**

The notion that the masses are too driven by base emotions, too irrational and hateful, too bigoted and brain-dead to be trusted to have more say in a civilised society has been around at least as long as the idea of democracy itself.

In ancient Athens, as we have seen, to denigrate democracy the great philosophers argued that the mass of citizens were asses led by pastry chefs who pandered to their base desires. The flipside of this argument was the need for an enlightened elite to rule, who could protect civilisation from the brutish mob.

The essence of that argument has recurred in one shape or another through history. But the precise shape changes with circumstances. The difference is that, in the modern era, the case against trusting power to the masses is not presented as a defence of privilege or property rights. Instead it is most often presented as a defence of human rights and especially minority rights, against 'the tyranny of the majority'.

It is no longer just aristocrats and frightened old-fashioned elitists who sneer at the majority as a potentially mindless mob.

Liberals are now often most likely to look down at the public in fear and loathing. If you give these people free rein, the argument goes, they will run wild and trample over the rights and civil liberties of others, taking us back to a dark age of lynching and oppression. Better, then, to leave these things to sophisticated judges and lords, or the experts who sit on quangos and run NGOs. They will defend human rights against the inhumane wrongs of the pro-hanging, anti-migrant majority.

Democracy is certainly a dangerous business, and the majority can indeed make unpredictable and unwelcome choices. Once you let the democratic genie out of the bottle, there is no guarantee that he will obey your commands or grant your wishes. A public debate about anything from capital punishment to immigration rules can go either way. If not, there would be no point having a debate in the first place.

There is no shortage of historical examples of where democracy has not gone the way that progressives and liberals might like. Some who fear giving democracy its head will point to more current events, such as right-wing politicians in the UK and the US who want to reverse laws on human rights or legalised abortion. And of course there is the Brexit referendum result, (wrongly) claimed as proof that popular racism is on the march in the UK.

So no, democracy does not come with any guarantees or extended warranties, and you can't always get what you want out of the ballot box. But you cannot pick and choose which bits of democracy you like, or who gets to enjoy the politics of choice, either. That would turn a liberty into a privilege to be dispensed to the deserving. And the alternative, of leaving it in the hands of a chosen few to make big decisions for the rest of us, is far more dangerous if we want to live in a free and democratic society.

The downside of democracy is that you have to try to win the arguments with a lot of people who don't agree with you. They may not necessarily all have the wisdom of King Solomon or Simon Cowell, or the expertise of Einstein or Angelina Jolie. They still get the vote, and get to give the finger to your beautifully crafted manifesto if they so choose.

There are some hard arguments to be won. But if we avoid those arguments, take the apparently easier option, and leave decisions about our future in the unaccountable hands of a few experts and officials, the consequences for our democracy will be dire.

What is really behind the 'they're all racists and hangmen' allegation anyway? It is a combination of a mistrust of humanity, and a lack of faith in the liberal case. It is saying that the majority of people really are the deplorables they are painted as, and that you cannot win a reasoned argument with them.

This is prejudice masquerading as enlightened thinking. Take racism. The idea that we live in a racist society, and that xenophobia is on the march across the UK, is a fiction of the elites' fearful imagination.

Because the fact is that the argument against racism has largely been won in Britain. We now live in a more tolerant society than ever before. Not because our rulers passed laws telling us to be tolerant, but because public opinion and the culture have changed through experience, interaction and argument. Shifting public attitudes to race in Britain are actually an advert for putting your faith in democracy and freedom of speech.

This goes for the Brexit referendum, too. Immigration was certainly a concern for many Leave voters. But surveys show it was not the main one for many, and that the overwhelming majority are against any attempt to discriminate against, let alone deport, migrants living in the UK.

There was a telling moment after the referendum, when the Conservative government announced a possible crackdown on firms employing migrant workers. This was a political stunt designed to connect with what the Tories imagined were the real concerns of Leave voters; although her government was pursuing Brexit, new prime minister Theresa May was of course a Remain supporter, and shared all the common prejudices about Leavers being bigots. The reaction, however, took the Tories by surprise; a furious public backlash against their reactionary proposal to target foreign employees that had them back-pedalling faster than Mrs May's kitten heels could carry her.

Even if it were true that racist ideas were on the rise, the only way to challenge them would be through open public debate. The idea that migrants are better left to the tender mercies of the British elites turns history on its head.

Racism has indeed been bad in the UK at times, so some might argue (as they have since the Brexit referendum) that the pendulum could swing back the other way and unleash another wave of prejudice. But when racism has reared its ugly head in the past, it has rarely been from the roots of society upwards. Instead past anti-immigrant feeling has been encouraged and even institutionalised from the top down, through laws and political campaigns.

The Tories are most often associated with laying the foundations for racism in Britain: from passing the first modern act to control black immigration in 1962 and running a notorious election campaign with the unofficial slogan 'If you want a nigger for a neighbour, vote Labour' in 1964, to Margaret Thatcher's warning about Britain being 'swamped' in 1978 or the May government's attack on overseas students. But the Labour Party, too, has done its fervent bit to foster anti-immigrant feeling as a response to public dissatisfaction down the years, from the Labour government imposing virginity tests on Asian brides wishing to enter Britain in the 1970s, to the official Labour campaign coffee mug calling for 'Controls on Immigration' that sat cosily in party leader Ed Miliband's many kitchens during the 2015 general election campaign.

The notion that the political elite should be trusted to protect migrants against the masses looks like a bad case of putting the fox on door duty at the hen house. History suggests that equality and minority rights advance as part of the struggle for more democracy – not as the result of an elite backlash against it. The abolition of slavery in the US, for example, did not happen because a learned judge found against the Southern slave-owners in a court. Abolition was the result of a bloody civil war that completed America's national democratic revolution, and forged one nation where the will of the majority could prevail over sectional interests.

And what about hanging, the traditional example wheeled out as proof of the perils of allowing the public to decide important issues? Since MPs voted to abolish capital punishment for murder in 1965, the belief has often been expressed that British public opinion clamours for the return of the noose, and possibly public hangings, and that this popular barbarism is only held in check by the decency of MPs who keep a parliamentary stranglehold on the issue.

No doubt there remains sizeable support for the return of the death penalty. That does not make it an illegitimate subject of public debate, just because some find it offensive and distasteful. War kills far more than any hangman. Should we take that issue off the political table, too?

Indeed, as with racism, UK public opinion on capital punishment has shifted with the changing times and balance of forces. A British Social Attitudes survey of 2015 found that, for the first time, fewer than half of those surveyed supported the return of capital punishment; 48 per cent, down from 75 per cent in the first such survey in 1983.[1]

In the distant past there have of course been lynchings carried out by angry mobs in civilised Western nations, most notoriously of black men in the US. But the argument about capital punishment is not about some perverse 'right to lynch'. It is about the power and organised violence of the state, and whether the criminal justice system should have the right to carry out lawful executions. It is, as always, the state not the public that poses the biggest threat to liberties and civilised values.

This is a point of more than rhetorical interest. There are, or were the last time I looked, no mobs marching through the streets of London chanting, 'What do we want? String 'em up! When do we want it? Dawn!'

In contrast to this imaginary threat of public hangings at the O2 Arena, however, there has been a very real illiberal assault on civilised standards of justice being carried out by the UK authorities with apparent impunity and almost without protest.

What about the protection against 'double jeopardy', an historical safeguard to prevent the state repeatedly prosecuting/persecuting somebody until it gets the 'right' verdict? Established in Britain around 800 years ago, it was casually abolished by the New Labour government in its Criminal Justice Act 2003 – a backward step for criminal justice widely hailed as an advance for victims' rights.

What of jury trial? Often said to originate from one of the few clauses of Magna Carta that still have practical relevance to

our system of justice today, the right to trial by jury is recognised as the eternal 'lamp that shows that freedom lives'. Yet
Magna Carta's last leg appears to be living on borrowed time.
Formal and informal changes made to the system mean that,
according to Geoffrey Robertson QC, as many as 97 per cent of
criminal cases might now be decided by lay justices or district
judges sitting alone.[2] Jury trials have effectively already been
abolished in civil cases such as libel trials – leaving our freedom
of speech entirely in the hands of judges.

Perhaps we should not be surprised that there are so few
liberal protests against these state assaults on justice. After all,
if you don't think 'ordinary people' should be entrusted with
decision-making on big issues, why would you believe in trusting a crowd of a dozen randomly selected jurors to reach a
verdict in important criminal trials? Far better surely to leave it
to m'learned friends, the lawyers and judges. Indeed there are
now feminist demands to have judges sitting alone to try allegations of serious sexual offences, since 'juries have no place in
rape trials'.[3]

Yet the truth is that jury trial has been a fortress to defend
the liberties of the persecuted down the centuries. One of the
earliest cases which established the right of a jury of 'twelve
good men and true' to decide the truth came in 1670, when an
Old Bailey judge locked up an entire jury for two nights 'without food, water, fire, tobacco or chamber pot' because they
refused to obey his order to find two Quakers guilty of seditious
assembly. When they still refused to comply, the insolent jurors
were fined; two would not pay. They eventually won, the Lord
Chief Justice ruling that the jury members were the 'finders of
fact' in court and that the judge had no power to interfere in or
overturn their verdict, just because he had decided the defend-
ant was guilty.[4]

Of course the danger with a jury trial, as with a democratic debate, is that you might lose the argument and the case. But it at least gives us a chance of living in a freer and more open society, where the decisions people make really matter and make a difference.

If you want people to act 'responsibly', then first make them responsible for taking decisions about society so that they feel responsibility for what happens. It is certainly preferable to leaving our liberties to the tender mercies of the state authorities backed by illiberal liberals who appear to believe that most voters are a racist lynch-mob, and that jurors are the rapist's friends.

## 2 'People are too ignorant to know what's right – leave it to the experts'

What proved to be the most controversial statement during the EU referendum campaign in the UK? Not Boris Johnson's typically overblown comparison of the EU's expansionist aims with Hitler's Germany. Nor prime minister David Cameron's desperate claim that leaving the European Union could mean signing up for the Third World War.

No, the statement that appeared to cause most outrage in political and media circles was Leave campaigner and then-Tory cabinet minister Michael Gove's suggestion to a television interviewer that 'I think the people of this country have had enough of experts'. The UK's sizeable class of experts and officials, who had all been warning that disaster would follow a vote for Brexit, responded as if Gove had demanded that they should all be lined up against a wall and shot.

The outrage was apparently still boiling a couple of months later, when Lord Gus O'Donnell – former head of the UK government's civil service – was interviewed in *The Times*.

'Perhaps most offensive to Lord O'Donnell', noted the sympathetic interviewer, 'was Leave's reckless dismissal of expert advice.' O'Donnell the ex-mandarin seemed barely able to conceal his condescension as he pointed out to the rabble-rousing politicians that 'when your car goes wrong, you actually do want to take it to a mechanic – an expert on cars. I wish Mr Gove very good luck when his car goes wrong and he decides he's not going to have the experts involved.'[5]

The argument that experts are better placed than ordinary people to make informed decisions about important issues facing society is another age-old case for undermining democracy that has come back into fashion lately.

In ancient Greek times the philosophers argued openly for replacing democracy with the rule of 'the wise' over the ignorant masses. They believed that, as one recent historian has it, democracy 'gave people what they wanted from day to day, but it did nothing to make sure they wanted the right things'.[6] Appeals to the masses were deemed to be based on 'flattery and lies' rather than wisdom.

The ancients' honest elevation of the experts over the electorate sounds refreshingly frank compared to the sophistry we hear today. The philosopher Plato, through his written record of what his mentor Socrates was supposed to have said, offered an early and more explicit version of Lord O'Donnell's argument about fixing cars. Plato's Socrates had no truck with the notion that the populace should make decisions about the future of Athens, and no inhibitions about spelling out his feelings against democracy. After all, he observed, when it came to deciding on technical matters such as shipbuilding (or perhaps, in another life, car maintenance), the popular assembly of Athens would consult the experts, not the ignorant. So why turn to people who lacked expertise in politics when deciding political issues?

It was ridiculous, Socrates asserted, that 'when it is something to do with the government of the country that is to be debated, the man who gets up to advise them may be a builder or equally well a blacksmith or a shoemaker, a merchant or ship owner, rich or poor, of good family or none'. Nobody was willing to dismiss these lowly non-experts by pointing out 'that here is a man who, without any technical qualifications, unable to point to anybody as his teacher, is yet trying to give advice. The reason must be that they do not think this is a subject that can be taught.' Socrates by contrast believed that political judgements could and should be taught by experts, rather than left to the verdict of the vulgar and uneducated citizenry.[7]

These days those who would put experts on a political pedestal are rather less honest about their opinions of the people. Of course we believe in democracy, they will insist, and of course the electorate must have its say. But ... there are things about which ordinary people have little or no understanding. In these matters people should allow the experts to decide, or at least cast their votes according to advice/instructions given by those in the know.

For example, they might say, what does your average voter know about the setting of national or European interest rates for the international money markets? And should the politicians who might do what their voters want, rather than what the capitalist economy really needs, be left to decide?

No, better by far we are told to entrust the setting of interest rates to the independent experts in the Bank of England, the European Central Bank and the US Federal Reserve. Never mind that millions of people's lives will be directly affected by whatever the bankers decide to do about interest rates. Democracy is all very well in its place, but its place is apparently not at the financial top table with the City money-men and

Euro-bankers. The important thing about 'independent' central banks is that they are supposed to be independent of any political interference – otherwise known as democratic control.

The central banks were handed these powers by politicians trying to offload democratic accountability for economic decisions. It is a sign of how the Left has submitted to this dogma that in the UK, the ability to take 'independent' decisions on setting interest rates was granted to the Bank of England as the first act of Gordon Brown, seen as a left-wing chancellor of the exchequer, after Labour won the 1997 general election. Elsewhere in Europe the idea that electorates and elected politicians cannot be trusted to know what the economy needs has been used by the EU and the IMF effectively to usurp elected governments of Greece, Italy and Ireland and impose technocratic experts to run the country as the markets require.

There are many other areas we are told voters do not understand. Bring on the experts to inform and guide the process of government. Give power to unelected and unaccountable bodies, the courts, commissions and consultants, the auditors and official inquiries, to make informed decisions and proposals that are beyond the ken of the voters in the street. And of course, line up the experts and officials to lecture the electorate about how they must vote in the biggest political referendum in memory, or an election for leader of the free world – and then castigate millions of the feckless little citizens for failing to do as they are 'advised'.

Expert advisers have an important role to play in any democracy. But there are questions which any amateur should ask before accepting that these people know what's best for the rest of us.

First of all, why should those academics and experts be taken at their word, even on the issues about which they claim expertise? The word expert comes originally from the Latin verb meaning to test, try, find out, prove. They in turn should always be tried and tested to prove their expertise by experiment and experience.

This brings us back to Michael Gove's original remarks that caused such consternation. The full version of what the former cabinet minister said – whilst being constantly interrupted by the Sky News interviewer Faisal Islam – is worth digging out. 'I think the people of this country have had enough of experts from organisations with acronyms saying that … they know what is best and getting it consistently wrong because these people … Because these people are the same ones who have got it consistently wrong …'[8]

And that was as far as he ever got. But the meaning was clear enough. Not a dismissal of expertise in general, but of the specific 'experts from organisations with acronyms' – like for instance the EU, the ECB, the IMF, the CBI and many more – who were telling the British people they had to vote to Remain or face inevitable and imminent economic catastrophe.

As Gove observed, these were the same experts who 'have got it consistently wrong' about what was going to happen to the economy in the past. Not one of them, for example, correctly predicted the financial crisis that struck the West in 2008, a time when almost all of the experts were telling us the world economy was set fair for smooth sailing. (Those who, on the other hand, kept endlessly predicting a crash and recession should not get much credit either, since the history of capitalist boom-and-bust cycles suggests they were bound to be right eventually. Even a stopped clock is right twice a day, as they say.)

So having tried and tested these economic experts on the basis of their experience, why should anybody have swallowed their instructions without question on how to vote in the EU referendum? Expertise in an area is not the same thing as infallibility. These people and institutions were often making political statements about their own support for the EU, disguised as expert economic advice.

Second, if experts are meant to be the unimpeachable source of public wisdom, what is the public supposed to do when different experts disagree? There are relatively few pressing issues where all scientists or authorities will concur with one another about the causes and consequences of a problem. Is nuclear power, fracking or wind the answer to our future energy needs? Is screening for prostate cancer a help or a hindrance to men's health? It depends which expert you listen to. Even on an issue such as climate change where, we are often informed, 'the science is settled', there are well-informed sceptics who disagree with the consensus and point out that the scientific method means everything should always be open to question.

And what about the many occasions when the experts change their minds, and yesterday's heretical fringe nonsense becomes today's accepted wisdom, or vice versa? These flip-flops appear particularly common among the supposed experts on issues of public health, where the advice on what to eat and drink or how to care for our children seems to change as often as a baby's nappy. It often brings to mind the words of American author Mark Twain, who advised readers to be careful when reading health books as 'you may die of a misprint'.[9]

The problems of experts disagreeing and changing their opinions highlight the importance of people hearing all the evidence and making their own judgements about what they believe to be true. Rather than submitting to mind control by the experts, it is surely right to take advice and then make up our own minds as reasonably and rationally as possible. This is truest of all in politics. A trip to the polling booth should not be the same as a visit to the doctor; that's why we each get to cast an independent vote as we see fit rather than simply following the prescription issued by the men and women in white coats.

Most important, from Socrates in ancient Athens to Gus O'Donnell in *The Times* of London, this focus on the role of experts continually confuses technical expertise with political judgement.

The good Lord O'Donnell is correct, of course, in advising Mr Gove and anybody else to take their car to an expert mechanic when it is in need of repair. They are the people who know more than the rest of us about the inner workings of the internal combustion engine and how to fix one. (Even here, however, there might be more than one motorist whose bitter experience at the garage suggests that the terms 'car mechanic' and 'expert' are not always necessarily interchangeable.)

So yes, we generally go to a qualified expert for car repairs – or, in Socrates' examples, for advice about shipbuilding or engineering projects. It does not follow, however, that a car mechanic or an engineer should go to a supposed expert in economics or political science for guidance on how to vote. Their expertise is far more questionable than that of technical experts; as even Daniel Kahneman, winner of the Nobel Prize for Economics, suggests, 'in long-term political strategic forecasting, it's been shown that experts are just not better than a dice-throwing monkey'.[10]

And regardless of how well read and accomplished these experts might be, this is politics. It is about values, morals and judgement far more than bare facts and figures. It is about everybody taking part in a democratic debate, coming to their own decision, and making their choice about which side – if any – to support. Democracy is founded and survives on the assumption that every citizen is equally capable of making that choice for themselves, of playing an equal part in deciding our society's destiny on important issues. That is why the mechanic and the academic expert get one vote each.

(In a vestige of the old British rule denying the right to vote to 'criminals, lunatics and members of the House of Lords', Lord O'Donnell is barred from voting for an MP in parliamentary elections, but gets the same vote as everybody else in local and European elections – and in a referendum.)

Old Socrates' apparent disgust with those who 'do not think this is a subject that can be taught' shows that, even with the greatest minds, prejudice can overrule intelligence. Political judgement and decision-making is indeed something that cannot be taught. It can certainly be learned – through debate, experience and interaction with others. But not taught, by expert instructors feeding you facts and The Truth, as if they were explaining an instruction manual. Democracy must be about trusting the judgement of the electorate, the wisdom of the masses, even if there are no guarantees they will produce the result you want.

Politics is not about somebody's answer being right or wrong for everybody. It is not at all like mending a car, where suggesting that you simply whack the engine with a large hammer should rightly make your advice invalid. Politics is about clashing opinions and interpretations, where the advocate of the hammer has the same chance to persuade the public as the fan of the screwdriver.

If somebody counters that what we need are more 'informed choices and decisions', the question immediately arises: informed by whom? In pursuit of whose interests? Who believes in expert or 'fact-checking' angels, floating on a cloud above the morass of humanity below, with no agenda to pursue or axe to grind?

In democratic debate, people are free to reach their own conclusions. Even if the experts might have a point, nobody is under any compulsion to let it override their own beliefs. We can certainly turn to experts for information and facts. But how that information is interpreted, what meaning we assign to those facts – this is the business of politics. And politics is above all about making a choice.

So, for example, in a debate about Brexit and its impact on the economy, we could well accept that the UK might be in for a bumpy ride, at least in the short term, but still think that democracy and freedom from the EU was ultimately the most important consideration. It was not the case that Leave voters simply ignored the experts or swallowed the Leave campaign's lies. It was perfectly possible to interpret all the doom-laden forecasts we were offered about the economy, and still conclude that it would be well worth losing a bit of foreign investment or value off the pound in order to gain greater potential political control over the future of our communities.

It is worth looking back to Tocqueville who, nearly 200 years ago in his classic work *Democracy in America*, believed that a democratic society needed a good system of public education to thrive. What it did not need, however, was anybody acting like 'father figures', laying down the law to the people as if they were teachers lecturing a classroom of children. That could weaken the popular resolve to take responsibility for their own circumstances just as surely as an authoritarian government. Tocqueville said of American voters then, 'I do not fear that in their chiefs they will find tyrants, but rather schoolmasters.'[11] That is a warning to ring down the years to our age of expert would-be schoolmasters.

When all that is said and done, no doubt many people may be under- or ill-informed on some big issues, whether about Europe, foreign policy or the economy. That is not however an argument for less democracy, but for more.

Shortcomings in public knowledge about the intricacies of political issues are not a reflection of any shortfall in the intelligence of the public. It is the way that people have been shut out of public affairs which has made them lose interest in debates about, say, interest rates. Why bother struggling to understand issues which you are repeatedly told are none of your business?

Christopher Lasch, American author of *The Revolt of the Elites*, draws an enlightening contrast between the present-day stagnation and the robust democratic exchanges of past US politics, from the riveting 1858 debates about slavery and popular sovereignty between Abraham Lincoln and Stephen Douglas before the American civil war, to the tradition of town hall meetings which involved the citizens in arguing about public affairs rather more than once every four years.

Now America has the pseudo-politics of the over-rehearsed and empty presidential election debates. What Lasch calls 'the lost art of public debate' has left much of the electorate in the role of passive spectators:

Since the public no longer participates in debates on national issues, it has no reason to inform itself about civic affairs. It is the decay of public debate, not the school system (bad as it is) that makes the public ill-informed, notwithstanding the wonders of the age of information. When debate becomes a lost art, information, even though it may be readily available, makes little impression.[12]

What we could do with now to bring democracy back to life is not more lectures from political experts, but a renewal of the art of public debate, using 'the wonders of the age of information' to engage and argue rather than just for abuse. Dynamic democratic debate does need to be informed by genuine expertise and practical experience where relevant. That is not the same thing as being instructed and sanitised by the dead hand of self-appointed experts in what is good for us or which is the right direction in which to travel.

We might do well to recall the words of Moses Finley, the US-born classical scholar who came to the UK and became a Brit after being hounded out during America's anti-Communist witch-hunts of the 1950s. Imagining how he would respond to Plato's prejudice favouring a government of intellectuals and experts, Finley the ancient Greek expert put the case for a modern democracy in which the experts are under the direction of society, rather than the other way around: 'When I charter a vessel or buy passage on one, I leave it to the captain, the expert, to navigate it – but I decide where I want to go, not the captain.'[13]

## 3 'Globalisation means democracy and national sovereignty are outdated'

What use is it to talk about popular democracy, a system created in the tiny ancient city-state of Athens 2,500 years ago, as fitting a twenty-first-century world of global interconnectedness and uncertainty, where the push of a button in an Asian financial market can cause a currency crisis in Europe in the middle of the night?

This is a fashionable argument for diminishing the importance of democracy and national sovereignty, which has become increasingly influential in recent years from Western academia to the United Nations. There are two connected strands to its case.

One says that in a fast-changing interconnected world of globalisation and uncertainty, national democracies are increasingly impotent and out of date. They cannot control external events and cannot exercise real power. So we need international political action and global protest movements instead.

The other says that national sovereignty is no great loss anyway. It is a dangerous system based on the divisive old politics of nationalism. Better by far, then, to have international systems of global governance that can protect peace and rights – with political intervention and military force if necessary.

Globalisation is a word bandied about on all sides, apparently to mean whatever the speaker wants it to. But the overall impression is of a fast-whirling closely integrated world economy, where hidden market forces and complex financial algorithms move mountains of capital and debt in seconds and make the world go around. What can we hope to do against such forces, except maybe get on global satellite TV by staging a little riot outside a world economic summit?

Just lazily bandying the word 'globalisation' around as an explanation of everything does not necessarily make it true. How far is global economic integration today really qualitatively different from in the past? This is one of those areas where, contrary to the impression often given, the experts disagree. Some point out, for example, that world financial integration via global flows of capital is nothing new; the sums involved might be much larger today in quantitative terms, but integration now is not qualitatively different to the world economy on the eve of the First World War.

Indeed, '[i]n important ways,' observes American history Professor Carl Strikwerda, 'the pre-1914 world was more globalized than our world is today'. Top economic analyst Martin Wolf of the *Financial Times* agrees that 'almost all the widely cited aspects of contemporary globalisation are old. The difference is only one of degree, not of kind ... none of the following is new: the declining relevance of distance; the "ideas" economy; the "weightless" economy; liberation by "microchip"; "Jihad vs. McWorld" ... and the twilight of sovereignty.'[14]

And far from globalised finance being able to override national barriers at will today, cross-border economic activities are now arguably constrained by state actions far more than a century ago, through regulations, protectionism policies, non-tariff barriers and other controls, including on the 'free movement' of people.

None of this would have been news to one of the old masters of economic analysis, John Maynard Keynes. Writing in 1919, shortly after the end of the First World War, Keynes was already feeling nostalgic about the open world economy that had been lost when war broke out: 'What an extraordinary episode in the economic progress of man that age was which came to an end in August, 1914!' They had enjoyed a world economy in which 'the inhabitant of London could [while] sipping his morning tea in bed' purchase the 'products of the whole earth' or 'adventure his wealth in the natural resources and new enterprises of any quarter of the world', all by telephone; or he could get up and secure 'cheap and comfortable means of transit to any country or climate without passport or other formality ... and would consider himself greatly aggrieved and much surprised at the least interference'.[15]

For Keynes, state interference in the global movements of capital and capitalists only really began with the end of the open world economy at the start of the First World War. So much, we might think, for the modern notion that 'globalisation' has suddenly altered everything about the world economy and made it less possible for nation states to assert their sovereignty over the market.

It seems that much of the loosely defined talk around globalisation today is not really about economics and the new untrammelled power of global capitalism. It is more of a coded message about the powerlessness of people. Globalisation, like the Hand of God, is an invisible force apparently beyond our ken or control. The most powerful 'ism' of our age is not capitalism or (certainly not) socialism, but fatalism – the belief that people are powerless to shape their destiny in the face of external forces. In which case, what is the point of banging on about democracy?

The idea that we live in an age of unprecedented global change and international turmoil, putting events beyond the reach of people and nations, appears to be based on an ignorance of history rather than an expert understanding of the present.

The history of the past several centuries is an endless story of uncertainty, turmoil and chaos. Modern history has been, to cite one of the students in Alan Bennett's *The History Boys*, 'Just one f***ing thing after another!'

The emergence of industrial capitalism smashed the old ways into smithereens; as Marx and Engels examined in the *Communist Manifesto* of 1848, the new era was distinguished by 'Constant revolutionising of production, uninterrupted disturbance of all social conditions, everlasting uncertainty and agitation … All fixed, fast-frozen relations, with their train of ancient and venerable prejudices and opinions, are swept away, all new-formed ones become antiquated before they can ossify. All that is solid melts into air, all that is holy is profaned'.

Meanwhile capitalism was already expanding 'over the entire surface of the globe' in search of markets: 'It must nestle everywhere, settle everywhere, establish connexions everywhere.'[16]

The modern era of history has continued to be one of 'uninterrupted disturbance of all social conditions, everlasting uncertainty and agitation'. It has witnessed the biggest revolutions the world has ever seen, the bloodiest, costliest wars ever known to humanity, economic booms, crashes and disasters that dwarf anything which might have gone before.

Yet this long age of 'everlasting uncertainty' is also the modern age of democracy. It has been marked first by the re-emergence of the idea of democracy in Western politics, and then by the advance of democratic politics through struggles and setbacks. It has been through debating and getting to grips with the disruptive dynamic that people have been able to understand and to influence their circumstances, shaping the world in which we live.

Uncertainty is not the enemy. It has been the midwife to every dynamic leap forward – and step backwards – in history. It has always been hard to win and hold on to democratic rights in nation states, and to exercise any degree of control in a world where power is concentrated in other hands and bodies. But it is difficult to see why it should be considered harder now than it was in the past. Was it easy for the ill-equipped French labourers known as *sans culottes* – 'without breeches' – to overthrow the most powerful monarch in Europe? Or for women denied the vote or a voice to find ways to force the British establishment to concede their democratic rights?

In these uncertain times when there are few political solutions on offer it is surely more important than ever that we try to breathe new life into popular democracy. Change and uncertainty provide opportunities as well as problems. Let's not allow all the high-powered Global-degook to distort that and portray people as inevitably impotent today.

The second strand to the argument says that national sovereignty is not a good thing anyway, and no great loss. Sovereignty is seriously out of fashion. There is a whole school of academic thought promoting global governance and 'cosmopolitanism' as an alternative to national sovereignty. And powerful international institutions from the UN downwards have been putting these ideas into practice, acting against national governments through international courts, sanctions, NGOs and even military interventions.

These interventions by global powers in the affairs of developing nations are no longer seen as oppressive colonialism or imperialism attacking national sovereignty. On the contrary they are presented as a defence of rights within the target nation, global governance apparently acting on the side of the powerless against the powerful.

The problem is, however, that global governance inevitably means the denial of the possibility of democracy. And it is on democracy that liberties ultimately rest.

It is worth noting that, in the first instance, this new global governance can look a lot like old-fashioned great-power politics. Nobody is invading or prosecuting Western states. Indeed as David Bosco, author of *Rough Justice*, observed in 2013 after the first ten years of the International Criminal Court prosecuting crimes against humanity, 'for an institution with a global mission and an international staff, its focus has been very specific: After more than a decade, all eight investigations the court has opened have been in Africa. All the individuals indicted by the court – more than two dozen – have been African.' Meanwhile the ICC had shown no interest in investigating the bloody conflicts involving US, UK and Allied forces in Afghanistan, Iraq or elsewhere in the Middle East.[17]

Inevitably, however, the political war on national sovereignty also ultimately calls into question the sovereignty of Western nations themselves. Thus the advocates of closer European union openly argue that sovereignty should be shared, and that national borders are bad things.

Now, as an old internationalist, I harbour few nationalist sentiments beyond lukewarm support for the England football team. But I do still see it as necessary to defend the principle of national sovereignty against these contemporary attacks. Sovereignty, like any form of power, can certainly be used to pursue bad ends. Yet it at least offers the chance of democratic control.

The key connection here is between power and responsibility. National sovereignty offers people the possibility to fight to hold the powerful to account. There is no chance of that if power and authority is removed from the political arena and invested not only in unelected courts and other bodies at home, but also in the international judges and generals and NGO officials who impose their will on populations in the name of global governance. It might ostensibly be done on behalf of the oppressed, but its consequences are as anti-democratic as old-fashioned imperialism.

As Jeremy Rabkin, professor of government at Cornell University, wrote in 2000, in an unfashionable defence of national sovereignty:

> At the heart of sovereignty is the notion that power and responsibility must be linked ... When the law is bad or proves to have unforeseen consequences, it is important to know whom to blame – or whom to address when

seeking reform. Again, the idea of international governance points in exactly the opposite direction. Standards are set by distant authorities who may know little about the countries where they will be applied. And they don't really need to know because they are not responsible for the consequences.

Rabkin observed how the International Criminal Court:

> empowers an independent prosecutor to overturn national amnesty laws. Perhaps Russia would be better off if former communists were punished. Perhaps South Africa would be better off if former officials of the white supremacist government were punished. Perhaps Chile, Argentina, and other Latin countries would be better off if officials of previous dictatorships were punished. But these countries all decided it was more important to achieve a quick transition to democracy and strive for reconciliation between former enemies, even if that meant letting many guilty people go free. This was a decision that people in these countries had to make. Is it really better to have such decisions made by a distant bureaucrat? … The idea of international human rights points in the opposite direction – that it is not all that important whether particular policies are agreeable to the country that must live with them.[18]

Whatever its problems, national self-determination remains a fundamental democratic right worth fighting for. Just ask the Kurds battling ISIS.

On the radical wing of the anti-sovereignty movement are the anti-globalisation movements which claim that, with

national democracy outdated and all but useless in an interconnected world dominated by 'neo-liberal' market economics,
new globalised protest networks are needed.

According to one recent sympathetic account, the World
Social Forum and Occupy movements envisage a 'new democratic subject': a 'politically progressive cosmopolitan agent of
the neo-liberal age, able to move fluidly among sites of power
and resistance, from the national to the regional, the local and
the global ... The democratic subject is formed in the space
between local confrontation and global techniques of organisation and consciousness-raising', seeking 'a broader and more
fluid global community committed to democratic thought and
practices'.[19]

Blimey. But what can 'democratic thought and practices'
really mean once they are detached from the goal of popular
democracy in a sovereign nation? The mythical democratic
subject 'formed in the space between' the local and the global
seems to be floating around the globe (or at least the World Wide
Web) adrift from notions of accountability and responsibility.

Thus they can assume the 'democratic' right to speak on
behalf of the global masses, without any need to ask those
people what they might think or want. The famous slogan of
the Occupy movements that camped out in various Western
cities asserted, 'We are the 99 per cent!' A bold claim from
protests that did not even number anywhere near 1 per cent.
These 'new democratic subjects' have redefined democracy to
mean staging media-oriented protests 'on behalf of' the passive
populace, whether the people want it or not. They are speaking
for a 'new global civic society' that exists in their imagination
and web forums rather than on the streets.

As Ivan Krasten, author of *Democracy Disrupted*, observes,
'The protesting citizen wants change, but he rejects any form of

political representation. He longs for political community, but he refuses to be led by others. He is ready to take the risk of being beaten or even killed by the police, but he is afraid to take the risk of trusting any party or politician. He is dreaming of democracy, but he has lost faith in elections.'

What we are left with, as international relations lecturer Dr Tara McCormack summarises it in her review of Krasten, is 'a global movement of "venting": we update our Facebook status, or send an angry tweet, and then go back to our lives. It is a cry of frustration, a form of exit politics, that does not challenge the status quo … These protests … are a kind of performance of democracy.'[20]

What all the talk about globalisation and anti-globalisation has in common is a fatalistic view of democratic politics as futile or worse. The danger is that these arguments represent a self-fulfilling prophecy of anti-democratic doom. What we need instead is a positive vision of the possibility of greater democratic control that draws on the past gains of popular sovereignty, and looks to the future.

## 4 'Democracy leaves voters at the mercy of media lies'

It has become most commonplace in left–liberal circles in the UK and the US to hold the influence of the mass mainstream media responsible for the corruption of politics and the problem of democracy. The battle-cry 'I blame the meejah!' can be heard in debates about everything from climate change to Brexit or Donald Trump.

The implicit argument is that mass democracy is an illusion, because powerless voters cannot resist the power of the media. People are inevitably misled by the lies and myths peddled by big media outlets in pursuit of self-serving corporate and political agendas. It follows from this that in our democracy, despite the apparent electoral strength of the many, real influence often rests in a few hands at the top of media empires.

Since Britain's EU referendum vote, the latest incarnation of this argument has been to highlight the problem of what is called 'post-truth politics' – which is basically an attempt to say that Leave won because its lies outbid the truths of Remain in the popular mind. The referendum result was illegitimate, many argued afterwards, because it resulted from media misinformation and malice, which meant malleable people 'did not know what they were voting for'.

The attempt to blame the Brexit-supporting press for the Leave vote denigrates the voters as monkey-see-monkey-do idiots. It also ignores the fact that, while the tabloid *Sun*, *Daily Mail* and *Express* all backed Brexit, the readership of these newspapers today is far smaller than the reach of the BBC news outlets, which took a clear pro-Remain slant throughout the campaign (and afterwards). If anything, much of the media is now widely seen as part of the elite that people are revolting against, rather than as His Master's Voice instructing them how to vote.

After the US presidential election, leading Democrats sought to blame the media for giving Donald Trump too much publicity and thus enabling him to dupe voters, as if it might have been possible somehow to 'no platform' a presidential candidate and defeat him via the media.

To some of us, that 2016 election actually looked like a powerful argument against the blame-the-media outlook. After all, Trump was elected despite being generally vilified by the media throughout the campaign; as one critic summarised it, he won the popular vote in 30 of the 50 US states even though he had the support of only two major newspapers, the *Las Vegas Review-Journal* and the *Florida Times-Union*.[21] Yet millions of voters were simply not moved by the one-sided media coverage.

For others, however, this only proved that the American media did not use its supposed power over the people properly. In an article headed 'The Lapdogs of Democracy Who Didn't Bark at Trump', *Washington Post* columnist Dana Milbank insisted that, while reports suggested 91 per cent of media coverage had been hostile to Trump, that was still not enough: 'The problem is the media didn't show bias against Trump earlier and more often.'[22] Whether the media is accused of being too biased or not biased enough, the assumption remains that it can determine the outcome of a democratic election.

This is a good trick. It dresses up what is essentially an assault on democracy to make it look like an attack on corporate capitalism. In truth the criticism of Big Media is really aimed against the Big Public.

After all, the media could only exercise such a Pavlovian influence over the public, telling them how to vote or what to buy, if the people really are gullible and ignorant enough to do as they are told. What is presented as a critique of the influence of the mass media is really a condemnation of the alleged idiocy of the masses.

Of course the media plays an important role in public and political life. But taken in its own terms, the notion that the media shapes reality can make little sense. In whatever form, from old-fashioned print newspapers to the latest trendy websites, the media reflects rather than dictates. The clue is in the name – it is a collective medium for transmitting information from one place to another. The media reflects public life, and helps to shape our perceptions of it. To blame the media for the state of the world it shows us seems as smart as smashing the mirror because you don't like the reflection it shows you. As a young Karl Marx wrote ironically in a German newspaper 175 years ago, a free press is about as responsible for the changing world that it reports and comments on 'as the astronomer's telescope is for the unceasing motion of the universe. Evil astronomy!'[23]

The nonsensical notion that the media rules the world can only be accepted because it reflects the low view in which the public are held. Attacking the 'popular' press and mass media has become a code for denigrating the populace and dumping on the masses, the dopes who are supposedly duped by it and often pay for the privilege.

This is a prejudice that has been around, if not quite since the days of ancient Greece, at least as long as the printing press, which was first introduced into England by William Caxton in 1476.

From the start the Crown and the authorities sought to control what could be printed – and, more importantly, read. Their determination to police the press reflected their wish to police what the untrustworthy people could legitimately think and speak about, to prevent the spread of 'wrong' ideas.

A system of Crown licensing ensured that nothing could legally be printed without the permission of the Star Chamber, a secretive court of judges and privy councillors. Any public criticism of the Crown or its ministers could be branded as seditious libel or treason. Writers expressing the wrong opinions in print could have their writing hand cut off, or perhaps the letters S and L (for 'Seditious Libeller') branded on both cheeks.

The intention was always to prevent the public being misled by the 'inflammatory' popular press. When the English monarchy was restored under Charles II in 1660, the king appointed Roger L'Estrange as his official censor, with an army of spies to seek out and destroy unlicensed printers. L'Estrange despised the very idea of printing new-fangled newspapers for the masses, because 'it makes the Multitude too Familiar with the Actions and Counsels of their Superiors ... and gives them, not only an itch, but a kind of Colourable License, to be meddling with the government'.[24]

L'Estrange published two official, government-approved newspapers to tell London's growing band of avid readers what to think. Any printer trying to suggest an alternative version of the news or publish 'a Colourable License to be meddling with the Government' was taking their life in their inky hands. As late as 1663, John Twyn became the last printer to be hanged, drawn and quartered at London's Tyburn – now Marble Arch – for publishing 'a seditious, poisonous and scandalous book' justifying the people's right to rebel against injustice.

By 1695 Crown licensing finally lapsed. But the Crown and ruling class remained determined to keep British subjects in ignorance and control what the public was allowed to know. One key issue in the fight for a free press was the right to report what was said and voted on in parliament. Well into the eighteenth century any reporting of parliament was banned. In the 1760s, as John Wilkes and his radical allies fought for the right of newspapers to report parliamentary proceedings as part of their campaign for more democracy, and disenfranchised Londoners rioted for 'Wilkes and Liberty!', Members of Parliament recoiled in horror from the idea of the public learning what they were up to from trouble-making newspapers.

As one MP said during the last attempt to crack down on press reports of parliament, by sending impudent printers and their supporters to the Tower, 'It is unfit that the people should be misled by printers and it is for their good that they should know nothing but came from the authority of this House'.[25] No doubt his argument against the people being 'misled' by the media would have gone down well with some in parliament today. Fortunately for us the people of London refused to accept 'that they should know nothing' even if it was 'for their [own] good'; 50,000 rioted outside parliament in 1771 to show their support for Wilkes and Co, and won the liberty of the press to report what the politicians said and did.

In the nineteenth century, as demands grew for the extension of the vote, the British authorities tried to keep the booming radical press out of the hands of the people by imposing stamp taxes, to make newspapers too expensive for working-class people. After the Peterloo Massacre of 1819, when troops attacked a crowd demanding electoral reform in Manchester, the government extended the 'tax on knowledge' to publications carrying political ideas as well as news, and increased punishments for those found guilty of publishing 'blasphemous and seditious libels' up to a maximum of fourteen years' banishment from Britain to a penal colony. These measures were explicitly aimed at curtailing the influence of newspapers and pamphlets that might 'excite hatred and contempt of the Government and holy religion'. The *Manchester Observer*, which coined the phrase 'Peterloo' (a caustic reference to the battle of Waterloo), was branded by the Home Office as 'the organ of the lower classes' which sought to 'inflame their minds'. The *Observer* soon sank under the crippling costs of prosecutions.

In the more modern era of mass communications since the twentieth century, it has become difficult for the authorities in any formally democratic state to justify a crackdown on the freedom of the media. Yet the underlying prejudice about the dangers of allowing the mass media to 'inflame the minds' of the ignorant masses has remained intact. Laying into the vulgarity and propaganda of the popular press became a coded way of flaying the vulgar and pliable populace.

By now the condemnation of the mass media was no longer confined to kings and their cabinet ministers. From the late nineteenth century to the Second World War, the most virulent attacks were often led by high-minded intellectuals and literati looking down in disgust at the newspapers – and their readers.

Typically, the German philosopher Frederick Nietzsche declared that 'the masses vomit their bile, and call it a newspaper', while the English author D. H. Lawrence suggested that schools should be closed to discourage people from reading those 'tissues of leprosy', popular books and newspapers. H. G. Wells, who thought the masses as alien as the Martian invaders in his novel *The War of the Worlds*, believed a popular newspaper was literally a 'poison rag' which pandered to and inflamed people's base emotions. Meanwhile, notes John Carey, author of *The Intellectuals and the Masses*, Adolf Hitler too was expressing his disgust at the moral degeneracy of the European masses: 'Like many English intellectuals, [Hitler] blamed this degeneracy on the mass media, deploring the poison spread among the masses by "gutter journalism" and "cinema bilge".'[26]

More recently still the question has become, how did 'popular' – as in the popular press – become a dirty word among those who consider themselves to be on the side of the populace, the people?

The modern Left, in Western politics and academia, has been in the front line of the culture war against the mass media – now incorporating television and the internet as well as traditional newspapers – since the Second World War. As I have noted before, Hume's Law of Inverse Proportion states that 'the less fulsome support Labour and the Left receive from voters, the more fierce their attacks on the mass media become; the less certain they are of the loyalty of working people, the more certain they become that the popular press is exerting malign influence.'[27]

All political parties on both sides of the Atlantic have at times been guilty of peddling a version of the it-was-the-media-wot-lost-it excuse for their own problems, but Labour and the US Democrats have often pushed it hardest. The Anglo-American Left has projected its disappointment with the masses onto the mass media. The less it trusts the populace, the more it blames the demagoguery of the popular media.

In the UK in recent years, this has resulted in a push for tighter regulation of the press – a proxy for tighter regulation of the public. The phone-hacking scandal at the *News of the World* provided the pretext for a crusade to sanitise the popular press, under the auspices of the Leveson Inquiry. In 2013 the leaders of all the UK's major parties did a deal with Hacked Off, the celebrity-fronted tabloid-bashing lobby, to recognise a new regulator by royal charter – the first state-backed system of press regulation in Britain since the abolition of Crown licensing more than 300 years ago. No national media organisation has yet signed up for a right royal thrashing.

The result of the EU referendum gave another boost to the crusade to tame the media and protect the public from itself. The Remain campaign, backed by almost all of the Left, could not accept that it might have lost because of the weakness of its own arguments. No, it had to be down to the power of the lying press. One of the main Hacked Off lobbyists was quick to insist that 'the Leave victory, which relied so heavily on falsehood, would not have been possible without the full-blooded commitment of the right-wing press'. He called on the government to 'deliver the sticks and carrots needed to make the system work' and 'initiate the process of protecting the British public from press abuse'.[28]

The 'sticks and carrots' in question are the post-Leveson legal measures, yet to be enforced at the time of writing, which would allow UK courts to impose exemplary fines on publishers who were not signed up to the state-backed regulator, and make them pay both sides' legal costs even if they won the case. As I said at the time, if that is a 'carrot' for the press, it is one shaped like a baseball bat with a six-inch nail banged through the end.

The real 'abuse' of the public in all of this is treating them as a pathetic blob in need of 'protecting' from words in the media – for their own good, of course. This age-old prejudice against the populace has found its latest incarnation in the post-referendum idea of 'post-truth politics'. It is just a new way of expressing the old snobbery about gormless and gullible bigoted voters, suggesting that the Leave campaign won only by shouting its evident lies louder and in more populist fashion than the earnest truths of the Remain camp. One solution proposed is apparently to have a new official body of independent 'fact-checkers' policing political debates for dodgy statements.

The poisonous notion of 'post-truth politics' not only reinforces the idea of the electorate as too hopelessly ignorant and easily led to make rational choices on their own behalf. It goes even further; as *Spiked* editor Brendan O'Neill argues, 'It makes politics like religion'. The Remainer Left are the keepers of The Truth, and anybody who disagrees with them can be dismissed as 'deniers' and heretics.[29]

But political debate is not about the worship of some transcendental truth. It is the process by which we as a democratic society decide what we believe to be true at any time. To be effective and representative that requires a no-holds-barred debate, open to all-comers, with the cut and thrust of claims and counterclaims and controversy. Not a sanitised pseudo-debate policed by fact-checkers claiming to be watching from a non-existent neutral cloud above the fray.

The result may not always be as we would like. But give me the chance to engage in democratic debate any day, rather than have the truth decided on our behalf by those who, like the censors of old, want to protect the public from dangerous ideas 'for their good'.

A key element of that debate is freedom of the press – understood in its widest sense today, incorporating the web and everything else. It was Thomas Jefferson, one of the USA's Founding Fathers, who said that he would rather live in a world of newspapers without government than government without newspapers, so important was freedom of the press. That spirit of public debate as the lifeblood of political life remains key to revitalising democracy.

The irony is that the internet provides unprecedented opportunities for extending the freedom and openness of democratic debate about the future of our societies. If those still expending their time and energy shouting 'I blame the media' from the virtual rooftops stopped making excuses and devoted as much effort to creating an alternative media and new forums for debate, who knows what might result?

As ever, the big question remains – who do we trust to judge what is in the public interest to be published, read and listened to? Judges, government ministers and the elitist police of post-truth politics? Or should we trust the public to take in everything, make its own choices, and take responsibility for shaping the public world in which we live?

# Epilogue

# Spelling out the meaning of democratic freedom

What's the trouble with representative democracy today? It is not representative or democratic enough. Contrary to recent reports from Europe and America, democracy is not somehow running wild in a hate-filled frenzy of 'mob rule'. We have far too little democratic freedom, not too much.

Western democracy is already constrained and tamed to within an inch of its life. Power is invested in unelected state bodies such as Supreme Courts and EU institutions, or is exercised from on high by an insulated political class. And when people try to kick back against their lack of any real say in public life, the top-down solution proposed is to put popular democracy on an even shorter leash.

Yet the only way out of the West's political crisis is via more democratic debate and participation, not less. To make living in a democracy more meaningful, and to have an open argument about the controversial issues that divide us.

So, what is to be done? Which is the way to replace the demockracy we experience today with a more meaningful democratic system? How might we set about breathing new life into the ailing body politic and putting the *demos* – the people – back at the heart of democracy?

As we have seen, the modern history of representative democracy has been a constant struggle. The concern of the political and social elites has always been to try to separate the representatives at the top from the roots of democracy; in terms of the original Greek meaning of democracy they have sought to insulate their *kratos*, power or control, from the *demos*, the people. In recent decades, that is a struggle they have been winning, but at huge cost.

There is a need for something new that can draw on the best traditions of democratic politics and bring them to life for the twenty-first century. In short, if we cannot return to direct Athenian-style direct democracy and won't put up with this unrepresentative demockracy, we need to find a modern way to make the system more representative and more democratic; to give the public a more direct role in public life, whether that means in choosing their representatives, campaigning for change, or using the internet to do something that nobody has even imagined yet.

There is no ready-made model for creating a more democratic society today. As always, that will have to come from the people themselves, not from experts thinking up political experiments. What is required first is a culture change to revitalise the spirit of democracy.

Putting the *demos* back into democracy might mean putting up with some popular decisions you don't much like. But the current alternative, of increasing *cratos* – control – without the *demos*, is one no self-respecting democrat could vote for.

The first thing to do is start an argument for the unconditional defence and extension of democracy, to try to infuse politics with a more democratic spirit. We can begin by spelling out some basic ideas about Democratic Freedom that are too often forgotten in a world where everybody in authority pays lip service to democracy in principle while trying to deny it in practice.

**D is for DIRECT DEMOCRACY** – let's revive the spirit of Athens, minus the slavery and misogyny. The system of direct democracy practised by the ancient Athenians in the fifth and fourth centuries BC gave power to the all-male citizenry in a way unseen before – or since.

Through the Assembly, any adult male citizen could turn up to debate and vote on the rules governing their society. In the people's courts, citizens selected to serve as jurors by the drawing of lots could not only decide individual disputes, but make laws that framed Athenian civilisation.

This was a culture infused with the spirit of democracy and public debate, where the people's will was paramount and uncertainty and rapid change were accepted as the norm.

Such an ancient system could hardly work today. We are no longer dealing with a city-state that constitutes a few thousand active citizens. And thankfully democratic rights are no longer confined to privileged male citizens freed by the work of disenfranchised women and an army of slaves.

But the spirit of direct democracy and popular sovereignty is surely one we can take from ancient Athens and breathe into our culture today – updated for a free society where the universal franchise is as much taken for granted as women's equality and the abolition of slavery. And we can stand up for the legacy of direct democracy against its modern detractors, by defending unfashionable causes such as referendum results and trial by jury.

**E is for ENFRANCHISEMENT** – make it mean something more. After a long struggle, the universal franchise has been won in the West. Every adult now has the right to vote. No serious politician suggests taking the franchise away today, and some want to extend it to younger teenagers.

Yet what does it really mean to be enfranchised, in a system where voting often appears reduced to an empty ritual and the public play little role in political life between election days?

In its origin, the word 'enfranchisement' comes from the Old French for freedom and sovereignty. It once meant being freed from slavery. But not being a slave is not enough. To be truly enfranchised today, we need the freedom to act as sovereign citizens with the ability to decide our own destinies. Making enfranchisement mean more than the freedom to vote for the candidate you dislike the least every few years is a precondition for bringing democracy to life. That means people having more opportunity to make choices that matter, and take responsibility for the results.

M is for the MASSES – the one group it now seems legitimate to hate. The Brexit referendum result in the UK sparked an outpouring of bile from high places, vomiting down on to the 'ignorant', 'racist' and 'hateful' mass of Leave voters. The election of Donald Trump in the US prompted another explosion of masses-bashing, with Trump voters depicted as a 'low-information' mob of 'deplorables' marching to the polls brandishing hate-posting smartphones rather than pitchforks.

If there is a hatred on open display in Western politics today, it is the elites' increasingly overt fear and loathing of the masses, and contempt for the democracy that gives the *demos* – the people – the voice to speak for themselves.

Yet the 'revolting' masses remain the best hope for humanity. When it comes to making decisions about the future the expansive wisdom of the crowd outweighs the one-eyed view of the experts, bringing to bear real experience of life as it is lived in our society rather than the rarefied outlook of cut-off elitist cliques. And despite the revival of fears about 'the tyranny of the majority', mass democracy rather than state power still represents the best potential defence of civil liberties and individual rights. How else do people imagine those were wrested from the state in the first place?

O is for OUT of court – where democratic politics ought to stay. There is a dangerous trend today for spineless elected politicians to outsource authority over major issues to unelected judges in the UK or EU. Such political questions ought to be ruled out of court in a democratic society.

The 1998 Human Rights Act, which incorporated the European Convention on Human Rights into UK law, has handed judges the power to poke their prodnoses further into issues of public importance such as free speech and freedom of the press, issuing injunctions and rules on what news might be fit for us to read or hear. A plethora of public inquiries has allowed British judges to assume authority to decide important political issues.

These judicial chickens came home to roost in October 2016, when three high court judges saw fit to overrule the express wishes of 17.4 million Leave voters and tell the elected government it could not trigger Brexit without the permission of MPs and Lords in parliament.

We are forever being warned by the elites about the importance of preserving judicial independence. Nobody wants to see judges acting as government stooges in a free society. However, that precious 'independence' from political interference also means the unelected judiciary is free from any democratic accountability, answerable only to their own gods, consciences and more senior judges.

Which is fine if the judges are presiding over ordinary legal cases involving individuals in criminal trials or civil cases. But not if they seek to sit in judgement on political issues affecting the whole of society.

These judges are not angels in wigs, floating on high above the fray below, supposedly 'independent' of any agendas or axe-grinding. We should keep politics out of court, and judges' rules out of issues that need the widest public debate. If it please m'lud, or even if it doesn't, hands off our hard-won democratic freedoms.

C is for the CLERISY – the new enemies of democracy. In the past democracy faced a frontal assault from its avowed enemies – would-be dictators, ambitious army generals, kings and cardinals, who all believed in One Man, One Vote, just so long as they were the One Man who got to decide.

Things are no longer so straightforward. Now the more insidious challenge to popular democracy comes from elected politicians, liberal intellectuals and enlightened officials. These people are all believers in 'Democracy, but …' just so long as it does not involve granting too much power to the *demos*. Together they make up a class which American writer Joel Kotkin calls 'the New Clerisy' – an educated elite self-appointed to play a priestly role in society, 'serving as the key organs of enforced conformity, distilling truth for the masses, seeking to regulate speech and indoctrinate youth'. The Clerisy are anti-democrats who 'believe that power should rest not with the will of the common man or that of the plutocrats, but with credentialed "experts" whether operating in Washington, Brussels or the United Nations.'[1]

These are the new enemies of democracy. They are not goose-stepping about in uniforms but speaking softly and carrying a big intellectual stick to beat us into 'enforced conformity'. The New Clerisy are more dangerous to Western democracy than any old-fashioned fascist today.

R is for a RIGHT, not a privilege. At the start of the modern age, the vote in Britain and elsewhere was a privilege, not a right, extended only to those deemed deserving because they owned sufficient property.

Under pressure from a democracy-hungry people, the UK franchise was extended in piecemeal fashion through the nineteenth and early twentieth centuries. It was not until 1928, when universal adult suffrage was finally written into British law, that the vote became a right.

Now, bizarrely, there are influential modern voices demanding that the vote be treated as a privilege once more, in effect if not necessarily in law. There are the campaigners complaining that Brexit or Trump voters were 'too ignorant' or just 'too stupid' to be allowed to make such decisions. Top intellectuals and officials seriously suggest that we have 'too much' democracy for our own good. Meanwhile prestigious Western academics propose getting rid of elections, or giving greater weight to the votes of the educated and enlightened (such as university professors, by any chance?).

But democracy is a right for all, not a privilege for the few. You do not get to pick and choose which parts of the democratic process you approve of, or only recognise the results that go your way. The freedom to participate and vote in a democracy as a sovereign, autonomous adult must always involve the right to make the 'wrong' choices. Once you try to cherry-pick which bits of democracy you accept, the question is always: who gets to draw the line?

A is for ACCOUNTABILITY, without which authority is just authoritarian. The test of any democratic system is that those in positions of power and authority must ultimately be accountable to the people over whom they rule. Far too much power today is invested in unelected and unaccountable institutions, from courts to commissions, quangos to public inquiries. And the juggernaut of state control is travelling further and faster away from democratic accountability.

The late Tony Benn was a rare Labour politician who, despite his state socialist tendencies, still had a feel for the spirit of democracy. Benn developed a useful list of 'little democratic questions' to ask anybody in authority: 'If one meets a powerful person – Adolf Hitler, Joe Stalin or Bill Gates – ask them five questions: "What power have you got? Where did you get it from? In whose interests do you exercise it? To whom are you accountable? And how can we get rid of you?" If you cannot get rid of the people who govern you, you do not live in a democratic system.'[2]

Try putting those questions on accountability to a Supreme Court judge on either side of the Atlantic, a UK regulator of press freedom empowered by a royal charter, or a European Commissioner making laws for the entire EU. Then judge for yourself how democratic is the system we live under today. No authority without accountability!

T is for TRUST – taking a chance on democracy.

In the end many of the great questions and doubts about democracy raised through history come down to one simple question: who do you trust?

Not when it involves expertise in fixing a car or judging cake-baking in a tent. But who do we trust when it comes to making decisions about the sort of society we share – how it is run, for whose benefit, and what values it lives by and fights for?

Should we put our faith in an enlightened elite of experts, government officials, judges and Eurocrats to know what is best for the rest? Or should we take a chance and trust the majority – otherwise known as ourselves and one another – to make their own choices and take those decisions?

Putting our trust in the people is not a matter of having blind faith. This is about politics, not religion. Democracy is a risky and potentially dangerous business because the outcomes are unknown. There are no certainties in a democratic debate, except for the certain knowledge that you can lose as well as win. But trusting ourselves and our fellow men and women remains by far our best bet.

Trusting in real democracy gives us the chance to win an argument. It means that the majority has to reason out the problem and arrive at a decision. It may not always come up with the answer we would like. But trust me, the alternative – of leaving our fates in the hands of a few – is a far more dangerous chance to take.

I is for INFANTILISING the electorate. There is a fashion for treating adult citizens as emotional infants, incapable of making rational and considered decisions for themselves. The flipside of this is the proposal to put older children on a par with adults, by lowering the voting age and allowing sixteen- and seventeen-year-olds to take part.

This is presented as a way to bring the energy and enthusiasm of youth into the political system. It is popular with those seeking to recruit for their losing cause, such as Remain campaigners and supporters of Scottish independence.

Yet why should passionate young people be any more engaged with our deadwood version of democratic politics than any other age group? Evidence suggests that they are even less so. In the EU referendum in the UK, for example, much was made of opinion polls that suggested those aged eighteen to twenty-four largely voted to Remain. Rather less was made of the figures which suggested only 36 per cent of that demographic had bothered to vote at all.

Young people, perhaps even more than others, need to see that democratic politics has something to offer them before they will make any commitment. Giving them the vote as a largely unwanted gift is unlikely to win much gratitude. Worse, it risks infantilising the electoral system and treating all voters like sixteen-year-olds, to be patronised and lectured.

Democracy is a grown-up business for adults making reasoned decisions. The voting age of eighteen might seem slightly arbitrary, as such age limits generally are. But it is as good as any. The priority for re-energising democracy should be to treat voters who have reached that righteous old age as autonomous adults with minds of their own.

C is for CHOICE – the thing that ultimately makes democracy worth voting for. Everybody in Western politics supports democracy in principle these days. But the legal right to vote is only one part of democracy. What are we voting for?

Most leading Western political parties have become mere election machines rather than popular movements, offering different brands of managerialism rather than distinct ideologies or outlook. In that case the act of voting can mean little more than choosing which bank to use or coffee to drink.

Democracy has to mean more than casting a vote. It must involve making a choice about the sort of society we want to see, and taking responsibility for that decision. Unless elections offer us such political choices, they risk being empty rituals. Making the system both more representative and more democratic requires political parties and candidates to give the electorate something worth voting for. Without meaningful choice, a vote means little more than a pencil cross in an empty box.

F is for FREE SPEECH – the lifeblood of democratic politics. Unfettered freedom of speech is, if anything, even more out of fashion than popular democracy. The threat to freedom of speech and open debate today is not jackbooted political censorship. It comes from a more insidious culture of conformism; a crusade against any idea or opinion deemed offensive or hateful, the crusaders' banner emblazoned with the command 'You-Can't-Say-That!'

The nonsensical notion that 'too much' free speech is a problem, and that people's thoughts and words must be policed to protect public safety, has been used to constrain public debate on controversial issues such as immigration and identity politics. Those who believe that banning an argument is the same as beating it have only succeeded in handing legitimacy to the views they try to suppress.

An open debate about everything is the only possible way to win the battle against objectionable ideas. The alternative is simply to 'No-platform' any opinion you find disagreeable – the recourse of anti-democrats down the centuries. Without the lifeblood of free speech, democracy will wither. Nowhere in the world is speech 'too free'. More freedom of thought and speech and no-opinions-barred debate is always the potential solution, never the problem.

Democratic debate is the way that society decides what it believes to be true. That process requires the maximum possible input of ideas and opinions. What it does not need is the intervention of official fact-checkers and thought policemen to tell us what The Truth might be.

R is for REASON – the reason why democracy works. The elite view of the electorate today is as an unreasonable mob, who don't know what is good for them or for a civilised society and make irrational decisions about everything from what they want to who they vote for. As one former Labour MP has it, the votes for Brexit and Trump proved that people are 'doubtless idiots' and we're living in 'the age of unreason'.[3]

This prejudice masquerading as wisdom turns reality on its head. The age of democracy has also been the age of reason, based on the ability of individuals to reason and decide for themselves rather than being blindly guided by God, or rather by His appointed kings and cardinals on earth.

As the philosopher Immanuel Kant wrote in 'What is Enlightenment?' (1784), at a time when the democratic age was dawning and great intellectuals believed in humanity: 'This enlightenment requires nothing but *freedom*', most importantly, 'freedom to make public use of one's reason in all matters', without being told 'Do not argue!' by those in authority. Kant observed that, then as now, 'We find restrictions on freedom everywhere', but insisted that 'the public use of one's reason must be free at all times, and this alone can bring enlightenment to mankind.'[4]

It was the Enlightenment belief in reason and humanity that led somebody like the English radical Tom Paine to lead the charge for democracy at the end of the eighteenth century. Paine described the impact of those world-changing revolutions in America and France, in awakening the desire for change, sweeping away the rubbish of the past, and allowing people to draw the reasonable belief that things need not be like this and a better world was possible. 'The present age,' he concluded, 'will hereafter merit to be called the Age of Reason ...'[5]

It is high time for a new age of reason to dawn, based on

Paine's spirit of greater freedom and democratic control and a belief in humanity's capacity to think for itself. In response to the rubbish about irrational, unreasonable voters today, let us insist instead, with the citizens of ancient Athens, that it is 'the way of wild beasts to be held subject to one another by force, but the duty of men ... to convince by reason.' Democracy is thus the highest expression of our humanity. It is only reasonable that we should demand more of it rather than less.

E is for EXPERTS – ask them to advise, then let the electorate decide. This is the age of technocratic politics, when we are told that enlightened experts – whether central bankers or international diplomats – are best placed to make the major decisions that will shape all our lives. After all, they tell us, issues today are too complex and the public too ill-informed to decide wisely. (Funny, though, how that thoroughly modern-sounding argument for diluting democracy has been around since ancient times.)

Time to put the technocrats back in their place – giving us advice, not instructions. The gap that matters here is not between the knowledge of the experts and the public, but between technical problems and democratic politics. The latter involves much more than data and algorithms or computer models. It is about making value judgements and moral choices.

Leaving it to the learned elites in the courts and the committee rooms to make the big decisions might seem the easier, safer option. But it risks reducing democracy to an empty ritual, by effectively disenfranchising and demobilising the public, and dissipating any popular spirit for political change and progress.

As Kant argued 230 years ago, the consequence of allowing others to think and decide on our behalf is to render adults as children:

It is so easy for others to set themselves up as guardians. It is so comfortable to be a minor. If I have a book that thinks for me, a pastor who acts as my conscience, a physician who prescribes my diet, and so on – then I have no need to exert myself. I have no need to think, if only I can pay; others will take care of that disagreeable business for me.[6]

We need experts as a source of information and insights, interpreted through their professional experience. We do not need them to tell us what to do with it. By all means let the experts inform the people and our political representatives of their own deepest thoughts. But don't let them presume to instruct us shallow folk in what we ought to think or do. Every citizen is their own expert when it comes to deciding which side of an argument they want to be on.

E is for ELECTORAL REFORM – why rearrange the deck-chairs on a foundering ship? There is little point simply tampering with the details of democratic government – by, say, introducing proportional representation or people's juries or any of the other fashionable changes proposed by experts and academics to fix the system. Some of these proposals might have merit in their own terms, but they are trying to find a technical solution to a political problem of democracy.

Revitalising democracy is not a job for political scientists and policy wonks who lock themselves away to think up new systems. What's most important is not the system but the spirit, the culture of democracy. If people want to participate they will find ways to make that happen. If they see no point in engaging in public life, then no tricky technical plan will fool them into doing so.

Take the proposals for proportional representation in the UK. There is little doubt that voting systems based on PR – where the outcome of elections match the actual way people voted – are more formally democratic, giving smaller parties representing minority opinions more of a chance. Yet for such a change to be meaningful there would need to be an explosion of newer parties and ideas, standing for clear political alternatives. If it was simply a question of recycling votes among the existing unrepresentative outfits, it would be little more truly democratic than the existing system.

Even a radical democratic proposal such as replacing the British monarchy with an elected president would arguably have little impact when elected politicians are held in such low esteem that many people would vote for the monarch anyway.

We should have a public discussion about alternative measures to improve the democratic system. But how democratic that system truly becomes will always be resolved by public attitudes to it, rather than expert technical tinkering. There is no answer to be found in moving the goalposts on a mud-flooded pitch, rearranging the deckchairs on a foundering ship, or any number of other metaphors. It is a political and cultural question of public spirit.

D is for DIVISIONS – the basis of universal democracy. There is a widespread assumption today that divisions are necessarily terrible in liberal societies, as revealed by the Brexit referendum in the UK and the election of Donald Trump in the US, and we should all want to heal them. But in a sense divisions are what democracy is about.

Democracy thrives where there is serious debate across a political dividing line, not a cosy conformist consensus. It is about a contest between clashing perspectives, a full-blooded battle of ideas, through which the majority can decide which way it wants to go. As with all battles, there are winners and losers, bitterness and bravado. That's democratic life. The alternative is to fix elections or perhaps censor one side, to create an illusion of unity, like those polls where the dictator scores 98 per cent.

The problem today is that society's divisions are often presented in terms of identity and demographics – between ethnic groups, say, or generations – rather than ideas and outlooks. They then become seen as fixed and immovable, often leading to exchanges of identity-based abuse rather than engaged opinionated debate. The solution is not to try to wish divisions away in empty calls for unity, but to redraw the dividing lines. Down with the internecine identity wars, bring on the divisive politics of choice.

O is for the OLIGARCHY – still dictating the language of anti-democracy. Western political thought was basically invented by leading Greek anti-democrats, in order to depict the power of the *demos* – the people – in ancient Athens as mob rule. From the first, democracy was a dirty word used to stoke fears by supporters of rule by an oligarchy – the most-propertied and powerful few.

Some 2,500 years later, it might seem that we have inherited from the ancient civilisations not the spirit of popular democracy but the anti-democratic outlook of the oligarchs. Referendums are still spoken of with horror in high places, sometimes sneered at as 'plebiscites' – from the Latin for the *plebs*, the mob of ancient Rome. 'Populism' is used as a boo-word, as if any politics based on an appeal to the populace – from the Latin for the people – must immediately be suspect. And the label 'demagogue' is brandished as the Greek oligarchs used it, to mean a dangerous rabble-rouser; its actual meaning is simply 'leader of the people', a dangerous idea to an oligarch, but hardly anathema to democracy.

It is high time we broke the oligarchy's historical hold over the language of politics, and made popular democracy a positive ambition to be fought for rather than feared and reviled.

M is for the MINDS and hearts of men and women – where democratic freedoms will live or die in the end. Even in the US, which enjoys the sort of written, court-enforced constitution and bill of rights which many reformers dream of seeing in the UK, freedom and democracy are not ultimately kept safe by the rulebook or any paper procedures.

As US Judge Learned Hand (a judge who understood democracy) told a wartime rally for freedom in New York's Central Park in 1944, it is wrong to invest 'false hopes' in courts and constitutions: 'Liberty lies in the hearts of men and women; when it dies there no constitution, no law, no court can save it … While it lies there it needs no constitution, no law, no court to save it.'[7]

Indeed those who appear most keen on citing the authority of constitutional law and convention are often keenest on using it to thwart the will of the people. We might think of it as summoning the democracy of the dead in order to entomb the living.

Breathing new life into democracy is not something to be decreed from on high. It will have to come from a modern renaissance of the democratic spirit in the life of our society. The way to begin is by maximising democratic debate about the issues that face us. That means reinventing the politics of choice through a struggle between competing visions of the future. It will mean posing the question of who we trust, whose authority we respect, and who the powerful are accountable to at the centre of every issue.

Next time an expert dreaming up alternatives 'consults' us as to what we think of representative democracy, perhaps we should offer the revolting opinion that we think it would be a very good idea …

# Notes

**Chapter 1: From Brexit to Trump: '… but some voters are more equal than others'**

1. Peter Orszag, *New Republic*, 14 September 2011
2. *The Times*, 25 November 2016
3. George Orwell, *Animal Farm: A Fairy Story* (Penguin Classics, 2000 edition), p. 97
4. *Daily Telegraph*, reporting research by the Centre for Social Justice and the Legatum Institute, 30 September 2016
5. Reported on Newsweek.com, 5 September 2016
6. *Private Eye*, no. 1422, 8–21 July 2016
7. Huffington Post, 10 November 2016
8. Jamelle Bouie, 'There's No Such Thing as a Good Trump Voter', *Slate*, 15 November 2016; Jason Brennan, 'Trump Won Because Voters Are Ignorant, Literally', *Foreign Policy*, 10 November 2016
9. @SavionWright, Twitter, 17 November 2016; Jane Onyanga-Omara, USATODAY.com, 9 November 2016
10. Yougov.co.uk, 17 November 2016
11. Sky News, 3 November 2016
12. *Daily Mail*, 15 September 2016
13. David Lammy on Twitter, 25 June 2016
14. BBC News Online, 24 June 2016
15. *Guardian*, 20 October 2016, 3 November 2016
16. *Sun*, 26 June 2016
17. *New European*, 19 August 2016
18. Jonathan Freedland, *New York Review of Books*, 18 August 2016
19. www.spiegel.de, 11 March 2013
20. Martin Kettle, *Guardian*, 28 October 2016
21. See, for example, the Editor's Note at the end of 'The Republicans' Dr Jekyll and Mr Trump Problem', Huffington Post, 13 September 2016: www.huffingtonpost.com/entry/donald-trump-script-teleprompter_us_57d8117fe4b0fbd4676bb6ee4

22. Newyorker.com, 9 November 2016
23. Huffington Post, 12 November 2016
24. Spiked-online.com, 14 November 2016
25. Benjamin Disraeli, *Sybil, or The Two Nations*, 1845, book 2, chapter 5
26. www.merriam-webster.com/dictionary/clerisy
27. *Daily Beast*, 1 October 2012
28. Cited *Daily Mail*, 23 June 2016, updated 4 July 2016
29. *Washington Post*, 'Why I Voted for Trump', washingtonpost.com/graphics/opinions/trump-supporters-why-vote/
30. *Guardian*, 8 November 2016
31. Independent.co.uk, 9 November 2016
32. *New Statesman*, 15–21 July 2016
33. Ibid., 29 July–11 August 2016
34. A. C. Grayling, *Liberty in the Age of Terror: A Defence of Civil Liberties and Enlightenment Values* (Bloomsbury, 2009)
35. *New European*, 23 August 2016
36. Ibid., 28 November 2016
37. BBC News Online, 1 September 2016
38. Ipsos MORI poll, May 2016
39. *Newsweek*, 1 November 2016
40. *Chicago Tribune*, 9 November 2016
41. *Foreign Policy*, 10 November 2016
42. Westernjournalism.com, 16 November 2016
43. *Nottingham Post*, 24 June 2016
44. *Sun*, 27 September 2016
45. ComRes poll, June 2016
46. ICM poll cited cityam.com, 21 August 2016
47. *Observer*, 23 October 2016
48. Spiked-online.com, 14 November 2016
49. CNN, 8 November 2016
50. Paul Krugman on Twitter, 9 November 2016
51. *Slate*, 15 November 2016
52. Scott Alexander, 'You Are Still Crying Wolf', slatestarcodex.com, 16 November 2016
53. Fox News, 11 November 2016
54. Jenée Desmond-Harris, Vox.com, 9 November 2016
55. Refinery29.com, 12 November 2016
56. Heraldscotland.co.uk, 7 September 2016
57. News.sky.com, 11 July 2016
58. *Observer*, 26 October 2016
59. 'Why the Government Believes that Voting to Remain in the European Union is the Best Decision for the UK', HM Government leaflet, April 2016

60. *Observer*, 28 August 2016
61. *Prospect Magazine*, 18 August 2016
62. ABC News, 12 November 2016
63. Reported in Christian Science Monitor, 13 November 2016
64. CBS News, 29 November 2016
65. Daily Kos, 17 November 2016
66. Mashable.com, 6 November 2016
67. *Sun*, 19 September 2016
68. Quoted in Yuval Levin, *The Great Debate: Edmund Burke, Thomas Paine and the Birth of Right and Left* (Basic Books, 2014), p. 201
69. Cited in the *Guardian*, 15 September 2016
70. Entertainment/guardianoffers.co.uk

## Chapter 2: Taking the *demos* out of democracy

1. George Orwell, *Politics and the English Language*, www.orwell.ru/library/essays/politics/english/e_polit/
2. Quoted in Paul Cartledge, *Democracy: A Life* (Oxford University Press, 2016), p. 306
3. David Van Reybrouck, *Against Elections: The Case for Democracy* (Bodley Head, 2016), p. 1
4. *The Times*, 14 September 2016
5. *Asian Times*, 11 September 2016
6. *Washington Post*, 15 September 2016
7. *Slate*, 11 September 2016
8. Christopher Lasch, *The Revolt of the Elites and the Betrayal of Democracy* (Norton, 1996 edition), pp. 28–9
9. BBC News Online, 30 January 2010
10. Alternet.org, 13 October 2004
11. Huffington Post, 11 April 2008
12. *New York Magazine*, 1 May 2016
13. Cited *New York Times*, 23 March 2004
14. Speech to Johns Hopkins University, Washington, 1 May 2014
15. Cited *The Times*, 21 October 2016
16. Van Reybrouck, *Against Elections*; Jason Brennan, *Against Democracy* (Princeton University Press, 2016)
17. Van Reybrouck, *Against Elections*, p. 124
18. Brennan, *Against Democracy*, pp. 4–5
19. *The National Interest*, nationalinterest.org, 6 September 2016
20. Aeon.co, 29 September 2016
21. *New York Times*, 7 May 2013
22. Brad Sheerman on Twitter, 15 July 2016; Fox News, 15 July 2016
23. Edmund Burke, *Reflections on the Revolution in France*, from *The Works of the Right Honourable Edmund Burke* (Holdsworth and Ball, 1834), vol. 1, p. 474; cited in Anthony Arblaster, 'Democratic

Society and Its Enemies', in Peter Burnell and Peter Calvert (eds), *The Resilience of Democracy* (Frank Cass, 1999), p. 43

24. Quotes from Bruno Waterfield, 'E-who? Politics Behind Closed Doors', Manifesto Club report 2008, manifestoclub.info/ wp-content/uploads/EU%20Essays.pdf

25. BBC News Online, 11 July 2015

26. For example see David D. Green, *Democratic Civilisation or Judicial Supremacy?* (Civitas, 2016)

27. A. Burgess, 'The Changing Character of Public Inquiries in the (Risk) Regulatory State', *British Politics*, vol. 6, issue 1, April 2011

28. *New Statesman*, 16 October 2012

29. Quoted in Frank Furedi, *The Politics of Fear* (Continuum Books, 2005), p. 35

30. Ivan Krastev, 'The Strange Death of the Liberal Consensus', *Journal of Democracy*, vol. 18, issue 4, 2007

31. Walter Lippmann, *Public Opinion* (FQ Classics, 2007 edition), p. 75

32. *The Australian*, 5 November 2011

33. OpenDemocracy, 30 June 2016

34. Jennifer Tolbert Roberts, *Athens on Trial: The Anti-Democratic Tradition in Western Thought* (Princeton, 1994), p. 13

35. Brennan, *Against Democracy*, pp. 7, 8 and 78

36. Cited in the *New Yorker*, 14 September 2011

37. Mick Hume, *Trigger Warning: Is the Fear of Being Offensive Killing Free Speech?* (William Collins, concise edition 2016), p. 23

38. Thucydides, *The Peloponnesian War*, translated by Rex Warner (Penguin, 1954), pp. 117, 119

39. Cited in David Runciman, *The Confidence Trap: A History of Democracy in Crisis from World War 1 to the Present* (Princeton University Press, 2013), p. 6

40. Niccolò Machiavelli, 'The Multitude is Wiser and More Constant than a Prince', from *Writings on Livy*, 1517

41. Cited in Runciman, *Confidence Trap*, pp. 12, 14

42. Orwell, *Animal Farm*, p. 113

43. Rollingstone.com, 27 June 2016

44. Quoted in Anthony Arblaster, *Democracy* (Open University Press, 3rd edition, 2002)

**Chapter 3: A short history of anti-democracy**

1. Arblaster, 'Democratic Society and Its Enemies', p. 34

2. Keith Graham, *The Battle of Democracy* (Wheatsheaf, 1986), p. 1

3. J. S. McClelland, *The Crowd and the Mob: From Plato to Canetti* (Unwin Hyman, 1989), pp. 1–2

4. Cartledge, *Democracy*, p. 1

# Notes

5. Tolbert Roberts, *Athens on Trial*, pp. 42, 43
6. Frank Furedi, *Authority: A Sociological History* (Cambridge, 2013), p. 44
7. Tolbert Roberts, *Athens on Trial*, p. 46
8. Cartledge, *Democracy*, p. 99
9. Ibid., p. 103
10. Tolbert Roberts, *Athens on Trial*, p. 8
11. Cited in Arblaster, 'Democratic Society and Its Enemies', p. 35
12. Paul Foot, *The Vote: How It Was Won and How It Was Undermined* (Bookmarks, 2012), p. 7
13. Ibid.
14. Ibid., p. 22
15. British Civil Wars Project, bcw-project.org/church-and-state/second-civil-war/putney-debates
16. Cited in Michael L. Morgan (ed.), *Classics of Moral and Political Theory* (Hackett Publishing, 5th edition, 2011), p. 650
17. Cartledge, *Democracy*, p. 287
18. Spinoza, *Theological-Political Treatise*, Cambridge Texts in the History of Philosophy, cocodrilo.synaptium.net/wp-content/uploads/2013/08/spinoza-theological-political-treatise.pdf
19. Cited in Arblaster, 'Democratic Society and Its Enemies', p. 36
20. Cited in Tolbert Roberts, *Athens on Trial*, p. 181
21. Cited in Malcolm P. Sharp, 'The Classical American Doctrine of "The Separation of Powers"', *University of Chicago Law Review*, vol. 2, issue 3, 1935, chicagounbound.uchicago.edu/uclrev/vol2/iss3/2
22. Christopher Hitchens, *Thomas Paine's Rights of Man* (Atlantic Books, 2007), p. 28
23. Cited in ibid., pp. 73–4
24. Cited in Foot, *The Vote*, p. 75
25. Foot, *The Vote*, p. 80
26. Victor Hugo speech translated by Bruno Waterfield, from www2.assemblee-nationale.fr/decouvrir-l-assemblee/histoire/grands-moments-d-eloquence/victor-hugo-21-mai-1850
27. Nicole Etcheson, 'A Living, Creeping Lie', *Journal of the Abraham Lincoln Association*, vol. 29, issue 2, Summer 2008
28. *The Chartist Circular*, 1841, vols 1 and 2, p. 29
29. George Eliot, *Felix Holt, the Radical* (1866) (Penguin, 1995), p. 130
30. Cited in Arblaster, 'Democratic Society and Its Enemies', p. 38
31. J. S. Mill, *Autobiography* (Oxford University Press, 1971 edition), p. 153
32. Cited in Dennis Smith, *Capitalist Democracy on Trial* (Routledge, 1990), p. 26

33. Carl Schorske, *Fin-de-siècle Vienna: Politics and Culture* (Cambridge, 1981), p. 145
34. Smith, *Capitalist Democracy on Trial*, p. 6
35. Cited in Foot, *The Vote*, p. 175
36. Ibid., p. 218
37. E. Sylvia Pankhurst (author) and Kathryn Dodd (ed.), *A Sylvia Pankhurst Reader* (Manchester University Press, 1993), p. 74
38. A. Hitler, *Hitler's Table Talk: 1941–44* (London, 1953), p. 497
39. www.goodreads.com/author/quotes/7805.H._L._Mencken
40. Beatrice Webb, *My Apprenticeship* (Cambridge University, revised edition, 1980), p. 173
41. Cited in Runciman, *Confidence Trap*, pp. 104, 106
42. Ibid., p. 306
43. Francis Fukuyama, *The End of History and the Last Man*, 1992 (Penguin, 2012)
44. *The Writings of the Young Karl Marx, Philosophical and Social*, translated by L. Easton and K. Guddat (Garden City, NY, 1967), p. 206

## Chapter 4: For Europe – against the EU

1. Miguel Herrero de Minon, 'Europe's Non-Existent Body Politic', in de Minon and G. Leicester, *Europe: A Time for Pragmatism* (European Policy Forum, 1996), pp. 1–5
2. E. Hobsbawm, 'An Afterword: European Union at the End of the Century', in J. Klausen and L. Tilley (eds), *European Integration in Social and Historical Perspective* (Rowman and Littlefield, 1997), p. 268
3. Cited in Levin, *The Great Debate*, p. 200
4. Cited in James Heartfield, *The European Union and the End of Politics* (Zero Books, 2013), p. 5
5. Ibid., p. 7
6. *The Times*, 24 May 2016
7. Jan-Werner Müller, 'Should the EU Protect Democracy and the Rule of Law inside Member States?', *European Law Journal*, vol. 21, issue 2, March 2015, pp. 41–60
8. 'EU Policy and Destiny: A Challenge for Anthropology', *Anthropology Today*, vol. 21, issue 1, February 2005
9. Jan-Werner Müller, *Contesting Democracy: Political Ideas in Twentieth-Century Europe* (Yale University Press, 2012), pp. 40, 128
10. Cited in Heartfield, *The European Union and the End of Politics*, p. 72
11. Cited in internationaldemocracywatch.org/index.php/european-union